THE PHYSICS OF
SKIING

SKIING AT THE TRIPLE POINT

D0089586

THE PHYSICS OF

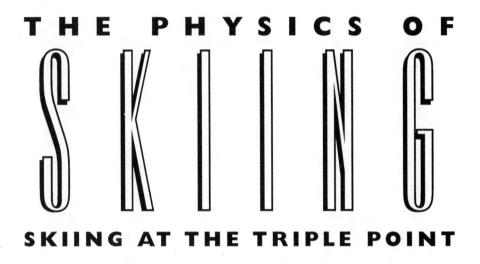

SKIING

SKIING AT THE TRIPLE POINT

DAVID LIND
Department of Physics
University of Colorado
Boulder, Colorado

SCOTT P. SANDERS
Department of English Language and Literature
University of New Mexico
Albuquerque, New Mexico

AIP
PRESS

Springer

Library of Congress Cataloging-in-Publication Data
Lind, David.
The physics of skiing: skiing at the triple point / David Lind, Scott P. Sanders.
    p.   cm.
    Includes bibliographical references and index.
    ISBN 1-56396-319-1
    1. Physics.   2. Skis and skiing.   3. Force and energy.   4. Friction.   I. Sanders,
Scott Patrick.   II. Title.
QC26.L56   1996   96-26220
796.93´01´531—dc20   CIP

Printed on acid-free paper.

Printed and bound by United Book Press, Inc., Baltimore, MD.
Printed in the United States of America.

9 8 7 6 5 4

ISBN 1-56396-319-1 Springer-Verlag New York Berlin Heidelberg SPIN 10691609

*Dedicated to the memory of Fred Lind,
who took a boy skiing for the first time
and started this journey.*

CONTENTS

Chapter 4

Alpine Skiing Techniques: Gliding, Wedging, and Carving ... 76

Chapter 5

Interactive Dynamics of Alpine Maneuvers 109

PREFACE

This is a book about skiing and a book about science. Readers who have some background in the physical sciences or in engineering should be able to appreciate the full discussion. For readers who are more skiers than they are scientists or engineers, we have tried to couch the discussions of technical matters in the main chapters so that they may be read and appreciated. Technical argument, however, risks losing its thrust unless some quantitative examples are provided. Wherever possible, the necessary quantitative examples are presented in simple mathematical relations. For readers who seek more fully developed technical expositions, a separate section of *Technotes* (or technical notes) offers short, augmented technical discussions of specific points made in some of the main chapters. In most cases, the technotes have much greater and more sophisticated mathematical and technical content than do the discussions found in the main chapters.

In part because this is a book about skiing and a book about science, it has two authors. Dave Lind is a scientist and a skier and a writer. Scott P. Sanders is a skier and a writer; he is no scientist. Allow us a few paragraphs to write in the first person singular and tell you something about the genesis of this book.

First, Dave Lind.

One of my first memories is of the walkway in the backyard of our house in Seattle in the early 1920s where I walked as a toddler, marvelling at snow up to the level of my head. I was not yet a teenager when, in 1929, my uncle, Fred Lind, took me skiing for the first time. I remember soon after that trip going to the local lumber yard to buy a pair of hickory boards from which to hand-fashion a pair of skis. In that place and in those days the only skiing was backcountry skiing: there were few lifts and no groomed slopes. In time I have become a reasonably proficient alpine skier, even, over the last 25 years, a competent free-heel or telemark skier. From my experience leading many parties on backcountry ski trips, I realized that there was a great deal of remarkable phenomena going on in skiing if one took the time to look for it, and I started to take that time.

I was trained as a physicist, first at the University of Washington and finally at the California Institute of Technology (Cal Tech), and I have taught throughout my career at the university level, retiring from the Physics Department at the University of Colorado in 1983. About 1970, when the practice of offering a variety of courses designed to teach physics to general audiences was much in vogue, I created a short course on ''The Physics of Snow.'' The course would give University of Colorado students some appreciation of the physical basis (snow) underlying one of their more favored activities—skiing. That course had considerable success because, in addition to the expected ski enthusiasts, geologists, geographers, and water-resource engineers, as well as professional ski patrol members, avalanche forecasters, and ski instructors had a real need to know more about snow and, in fact, enrolled for the class. I taught the class a number of times over a span of more than two decades. To a great extent, the contents of this book developed from my teaching the physics of skiing as one application of the physics of snow.

And now, Scott.

I first skied in 1966 on a pair of 215-cm rental boards at the single slope offered by the Broadmoor Hotel in Colorado Springs, Colorado, where I was a freshman at Colorado College, lately arrived from Southern California. All I remember is that I fell down, constantly. I left Colorado College after that year, and the Broadmoor has since had the good sense to close their ski slope. My skiing career began again in the mid-1970s when, as a graduate student studying English at the University of Colorado, I started cross-country skiing. In Boulder, cross-country skiing meant skiing up a mining road or a trail or some combination thereof to get as high up into the mountains as one could go, and then either skiing back down the trail and the road, or, more likely, skiing down the open slopes above the tree line and through the trees for as far as one could safely go. A few years later, I met Margy Lind. And her father, Dave Lind.

On our first ski tour up Washington Gulch outside of Crested Butte, Colorado, my labrador retriever and I lagged behind as Margy skied ahead and Dave skied on still further ahead, returning every now and then to see how I was doing, and then skiing back up the trail. By the time we finally stopped, I was exhausted, and Dave had certainly skied more than double what I had with his up and back and up again approach to our ski tour. Eventually I tried alpine skiing again, and I married Margy. After some years of practice, I no longer fall down (much) when I'm alpine skiing, and I know well the four most dangerous words in the Rockies: ''C'mon Dad, it's easy!'' I come to the writing of this book entirely through my relationship with Dave Lind, my enjoyment of skiing, and my love of writing.

Every book represents the cumulative effort of many people. Dave Lind acknowledges the professional, collegial, and personal help of all who freely offered helpful discussion, advice, and information along the way. Especially he thanks Dr. George Twardokens of the University of Nevada, Reno, who coaches ski professionals and whose persistent questioning and willingness to argue technical points was invaluable. Special thanks also to Dr. Sam Colbeck of the U.S. Army Cold Regions Research and Engineering Laboratory for supplying much technical information and encouragement. Thanks also to the many people who were always available to discuss this project, especially Dr. Bard Glenne, Chief Designer for the K2 Corporation; Ron Garrett of Volant Ski Corporation; Lou Dawson, backcountry ski mountaineering guide; and Dr. Richard Armstrong of the University of Colorado. Marc Dorsey and Linda Crockett of the Professional Ski Instructors of America (PSIA) helped discover useful, hard to find, information. Jim Middleton and Tony Forrest of the PSIA and Wolverton Productions provided video footage of PSIA demonstration team members that helped illustrate several skiing maneuvers. Thanks to Bill Semann, a former ski racer now technical artist nonpareil, who worked tirelessly creating the line-art illustrations for this book from a variety of sources. Thanks also to Boulder Ski Deals of Boulder, Colorado for providing the skis that were tested in the laboratory and shop of the Nuclear Physics Laboratory of the University of Colorado, whom Dave Lind thanks for the administrative assistance they provided at all stages of this project. Thanks go also to the colleagues and friends who shared many ski tours with Dave during which questions about the physics of skiing were often topics for conversation. Dave especially thanks his friend and colleague, Dr. Kurt Gerstle, Professor Emeritus of Civil Engineering at the University of Colorado and skiing enthusiast; together they spent many long hours discussing technical matters while skiing back-mountain routes.

Scott P. Sanders thanks the administrative staff of the Department of English, University of New Mexico for their support. Scott especially thanks his family for making sure that he had something other than the physics of skiing to think about, and for allowing him to be absent from their lives too much as this project moved toward completion. Both Dave and Scott thank the editorial and production staffs of AIP Press, especially Sabine Kessler, who offered several helpful suggestions for taking our work and making it a book. And, finally, we thank Maria Taylor, Executive Editor of AIP Press, for her patience and persistence and for believing in this project enough to see it through to publication.

To our knowledge, no other book written for the general public or for the scientist or engineer has tried to relate so many of the fundamental aspects of skiing to basic physical principles. There are many articles that offer technical treatments of many of the points that we discuss in this book; however, most of these documents are very technical indeed, and they are often not very easy to find. In several cases, they are not published in the traditional literature of scholarship on snow and skiing that one might expect to find in a reasonably good university library. Some of the sources we have cited in the Bibliography were selected in part because they contain extensive citations to this large but hard-to-find body of technical literature. Readers wishing to access the primary historical or technical literature should refer to the Bibliography to help them locate such sources.

Finally, we know and celebrate the fact that no matter how much the physical analyses presented in the pages that follow may explain the many mysteries of skiing, some part of that mystery will always remain for each of us, because each of us makes skiing our own.

CHAPTER 1

INTRODUCTION
At the Triple Point

This volume has an enigmatic title. The concept of "skiing at the triple point" is probably the key to this book. In many sports the properties of the playing field are relatively fixed and unchanged, and they remain so during the course of the play. That is definitely not so in skiing. A peculiar circumstance of skiing that it shares with one of its near relatives, ice skating, is that skiing can only be done on a playing field whose basic physical properties change. When we ski, small changes in temperature make huge changes in the playing surface. For skiing, the playing surface is water, which in the course of a single downhill run may exist in all its phases: as a solid (ice, in the form that we call snow); as a liquid; and even in the form of water vapor, as a gas. We shall see that in many ways, skiing works best near 0 degrees Celsius (°C) or 32 degrees Fahrenheit (°F), which is roughly the temperature of the triple point of water. Thus we may say that we ski at the triple point—where the three possible states of water (solid, liquid, and vapor) coexist.

But there is another triple point that this book addresses. In all sports, better performance and greater enjoyment occur when we understand the cause–effect relationships that transform the physical actions we perform into consciously practiced techniques. We skiers know that it feels good to carve a smooth, parallel turn on freshly packed, powder snow or to wheel down-mountain in deep, untracked snow throwing up rooster tails of snow in our wake and leaving a magnificent track of linked turns on the slope. In this book, we examine the physics of the many forces and properties that come together in this sport to give us those good feelings. Our goal is to ski at what we might consider to be a second "triple point," at the point where our increased understanding of the "how" and the "why" of skiing joins with our experience of the "wow!" and then we know the fullest enjoyment of our sport.

We have felt that ultimate "wow"—skiing at that second triple point where physical understanding meets the simple joy of experience—many times, and even though we have long been fascinated by snow and by skiing, in the course of writing this book our enjoyment of skiing has increased. We hope that in the course of your reading this book, a similar increase in enjoyment will be your experience, too.

SKIING THROUGH THE AGES

Skiing has both a prehistory and a history. Norwegian pictographs (see Fig. 1.1) and the archaeological recovery of ancient ski fragments suggest that some form of skiing pursued as a mode of travel over snow dates from at least 4000 years ago. Skis were but one of the devices that evolved to enable travel on foot through soft snow. Snowshoes, in a variety of different forms, probably developed independently from skis, and they were perhaps more widely used.

In historical times, we know that the Carthaginian general Hannibal encountered snow avalanche hazards when his troops crossed the Alps to attack Rome during the Second Punic War (218–201 B.C.). While we do not know that Hannibal's troops used skis, they surely used some such method

FIGURE 1.1. Pictograph from Rödöy, Norway, circa 2000 B.C. (Courtesy Norwegian Ski Museum, Oslo.)

of travel to expedite their passage over the snows of the Alps [1]. The somewhat more definite history of skiing begins with the work of the Byzantine historian, Procopius (526–565? A.D.), who described gliding Finns (apparently using what we might recognize as skis) racing nongliding Finns (apparently using what we might recognize as snowshoes). In the year 800, Skadi is mentioned in Scandinavian mythology as the goddess of the ski, and from that time on various writings survive that mention the use of skis for travel and for military maneuvers. Finally, in Europe of the sixteenth and seventeenth centuries, written descriptions of skis and bindings appear in print along with sketches of those early skis [2].

In North America, early explorers, trappers, and settlers most likely used snowshoes to cross snowy mountains and plains; the written record of skiing in North America does not begin until about the middle of the nineteenth century [3]. Skiing was probably introduced by Nordic and German immigrants, who carried their knowledge of skiing (and perhaps their skis, too) with them to the New World. In early Colorado mountain history, there are numerous accounts of using skis for transportation. At the height of gold and silver mining in the Colorado Rockies during the 1870s and 1880s, substantial mountain communities would become isolated by the heavy winter snows. In these severe conditions, skis were an important mode of transportation. In particular, mail carriers used skis when they delivered the mail. At the Crested Butte Mountain Ski Resort in Colorado, the Al Johnson Memorial Up Mountain and Down Hill Race is held every year in late March to memorialize the mail runs that Johnson made over the mountains to bring mail to the mining communities in the region. The histories of some mining camps record recreational ski competitions with a variety of events [4]. Another mail carrier and sometime racer, John A. "Snowshoe" Thompson, was known in the California gold fields for his skiing—and snowshoeing—exploits [5]. J. L. Dyer records his late nineteenth century travel by ski in the mountains around Breckenridge, Colorado, where he was a minister of the Methodist Church [6].

Around the turn of the century, organized ski schools with recorded competitions were introduced to America by Norwegian immigrants. Ski jumping, the special passion of the Norwegians, also dates its American beginnings from this time. The first analysis and description of skiing was published in 1896 by Mathias Zdarsky (1874–1946), an Austrian who is known as the father of the alpine ski technique. Two years after the First World War, another Austrian, Hannes Schneider, started the first organized ski school that used a definite teaching protocol. As a young man, Schneider read and was greatly influenced by Zdarsky's instructional book, *Lilienfelder Skilauf Technik*. Schneider came to the United States in 1939 and set up his ski school in North Conway, New Hampshire [7]. His school was soon followed by many others, most of which were associated with different national styles of the increasingly popular sport: there were French ski schools, German ski schools, Swiss ski schools, Italian ski schools, and, eventually, American ski schools.

THREE CLASSES OF SKIING

The are many, many different variations within the realm of what we may consider to be skiing in its broadest sense, and these variations use many different types of equipment and many different techniques. For example, consider the monoski, which offers the skier a single board on which both feet are mounted together and point forward. The monoski is maneuvered down the hill using techniques much like those associated with skiing done on two skis, mounted in the traditional manner, one for each foot. Also, we occasionally see on the slopes the ski equivalent of the rollerblade or ice skate, the foot ski or snow skate, which usually extends no more than a matter of inches past the heel and toe of the boot, making it more an adaptation of the ski boot than of the ski itself. Even so, snow skating is still more recognizably a type of skiing than it is a type of ice skating. The most popular recent innovation in the world of skiing has to be the snowboard, the snow-sliding equivalent of the skateboard or surfboard. The snowboard is much wider than the monoski, and the boarder's feet are mounted fore and aft across the snowboard, rather than pointing forward as they do on the monoski or on traditional skis. Snowboarders may be seen on every ski hill where they are not banned (as they are, at this writing, at Taos, New Mexico, for example). The snowboard and the boarder (when not airborne) carve graceful turns in the snow in a manner that should interest the traditional skier who is curious about the physics of carving a turn on skis, but more on that subject later. Finally, consider the variety of equipment that permits disabled skiers to experience the thrill of negotiating the slopes, some standing, some sitting, some using outrigger skis mounted on their poles.

For the purposes of our discussion, we focus on fairly traditional skis and skiing, done on two skis using the common complement of equipment: boots, bindings, and poles. We group all of the skiing of this sort into three principal classes: alpine, nordic, and adventure skiing.

Alpine skiing encompasses the faster-paced skiing events that take place down the pitch of steep slopes: downhill, super giant slalom, giant slalom, and slalom. Typically, the alpine skier rides a lift to the top of a run and then skis down the slope. Most general recreational skiing is alpine skiing. One defining factor is the alpine skier's equipment. The alpine skier's boot is firmly attached to the ski at both the heel and the toe by a binding that, to minimize the risk of injury, releases only in the event of a hard fall. The attached heel makes it difficult for a skier wearing alpine equipment to cover any distance over flat terrain; climbing hills in alpine boots and bindings can only be done for short distances by sidestepping up the hill with the skis across the fall line. Thus one defining feature of alpine skiing is going downhill, usually as fast as one is able.

Nordic skiing includes a variety of techniques. Classical or diagonal track nordic skiers negotiate a more or less flat course by skiing in parallel tracks, propelling themselves by poling and kicking alternately with their poles and skis in a diagonal relationship to each other: the right pole stretches ahead as the left ski slides forward, the left pole stretches ahead as the right ski slides forward, and so forth. A recent innovation on this technique, developed in competitive nordic racing events, is ski skating, which uses a wide, prepared track in which the skier slides one ski diagonally outward on its edge while pushing off against it, as in a skating motion. Finally, ski jumping is a form of nordic skiing. One feature common to each of these very different nordic skiing pursuits that distinguishes them from alpine skiing is that the heel of the nordic skier's boot, whether the skier engages in nordic track skiing, ski skating, or ski jumping, is not attached to the ski by the binding.

The final class of skiing we will call *adventure skiing*, adopting a phrase coined by Paul Ramer to describe all types of remote backcountry and mountain ski travel [8]. The cover of this book suggests what adventure skiing can be. The skier on the cover photograph skis down the untracked slopes of Chimgan in the West Tien Shan Range of Uzbekistan. Adventure skiing combines aspects of the techniques and equipment used in both nordic and alpine skiing to create a hybrid, third class of skiing. Historically, adventure skiing evolved in Europe with the practice of ski touring from alpine hut to hut. These high-altitude ski tourers developed modified nordic equipment with bindings that could leave the heel of the boot free for climbing and general travel, but could also fix the boot to the ski, attaching the heel for alpine maneuvers going downhill.

Another aspect of adventure skiing involves free-heel, downhill skiing, usually called *telemarking* after the region in Norway where this characteristic turning technique originated. Telemarking or free-heeling was originally primarily done by the more adventurous backcountry ski tourers, which explains its close association with adventure skiers. Its popularity has grown so much, however, that today free-heelers may be found practicing their turns on the groomed slopes of ski resorts, possibly tuning up before venturing out into the backcountry, to travel hut to hut, tour up and down a steep trail, or hurtle down chutes of untracked snow whose inaccessible entrances they have reached by climbing, by snowcat, or even by helicopter.

There is a some overlap in the equipment and the techniques associated with these three classes of skiing; but, in general, the differences between them are significant enough to warrant using these designations as a system of classification to help guide us through our consideration of the physics of skiing.

SNOW: THE PLAYING FIELD

The physical basis, the science, needed to understand the sport of skiing lies in a number of subfields. In a logical sequence, the nature of the playing field comes first, and that is the subject of Chap. 2. We consider the formation of atmospheric snow and the metamorphism that occurs in the ground-cover snow as the flakes that accumulate to form the snowpack deform with changes in temperature and environment. In this chapter we also consider the molecular structure of water near the triple point—the temperature and pressure at which water exists simultaneously in each of its three phases, as a solid, as a liquid, and as a vapor. Finally, we consider the thermodynamics associated with the phase changes.

EQUIPMENT

As anyone who decides to own, rather than rent, even the most basic ski equipment quickly discovers, purchasing ski equipment represents a major investment. There are a dozen or so distinct classes of skis requiring perhaps three or four different types of boots and as many types of poles. Unfortunately, the logos and elaborate graphic designs on much modern ski equipment tend to distract the consumer from finding essential physical information that should be, but more often is not, readily available. Skiers should ask questions about more than just the length of the skis they use; ski width, sidecut, fore and aft body stiffness, torsional rigidity, vibration damping ability, and shovel conformation are also important considerations. Few ski manufacturers readily display all of this information, and fewer still ski salespersons or skiers can define, much less compare, these physical features of skis.

In Chap. 3, all of the features of ski equipment that can help skiers achieve optimal performance are defined and discussed. Much of the discussion takes a structural engineering approach, considering the stress−

strain properties of the materials from which skis, poles, boots, and bindings are fabricated. Having some understanding of the flexural and dynamic properties of skis, boots, and binding systems can help skiers understand how their equipment is designed to perform. With that knowledge, we may better match our equipment with our abilities and preferred techniques.

SKIING TECHNIQUE

Most sports evolve largely by trial and error as practitioners experiment with equipment and techniques. Skiing is no different, and there are few careful, quantitative analyses of the mechanical science—the physics—underlying the activity of skiing. Most of the attention in discussions of skiing technique is given to one maneuver: making turns, especially making carved turns.

In all of alpine, nordic, and adventure skiing, there are essentially two classes of turning techniques: steered turning that employs some form of controlled skidding, and carved turning. The underlying physics of both turning techniques may be described as a mechanical system in which the skier moves down a slope and picks up kinetic, or motional, energy, just as Newton's apple gained kinetic energy as it fell to the ground. That motional energy must be entirely dissipated when the skier arrives at the bottom of the hill and is standing still. A small part of the energy is dissipated as heat from the rubbing of the skis on the snow; much more is dissipated by the skis' cutting, grinding, and throwing snow out of their path during the descent. Ski racers want to get down the slope as quickly as possible, so they carve their turns as much as possible, which yields minimum energy dissipation. Recreational skiers, keeping their speed under control, carve the snow to create stylish turns and skid their skis at controlled intervals to control their speed by dissipating their kinetic energy.

In Chaps. 4–6, we describe in detail the physics of most skiing techniques, giving special attention to the various techniques associated with carving turns on both packed and unpacked snow. Each of these chapters requires some understanding of Newton's laws of motion and energy, which figure prominently in the chapters themselves and in the technotes associated with the discussions. In these chapters, the physical activity of skiing is most directly connected to the physical science that describes and explains how and why what happens, happens.

FROM TRACKS TO TREKS

Chapter 7 considers the physical properties of the equipment and techniques specifically designed for nordic track, cross-country, and adventure skiing. Nordic track skiing is exclusively done on a prepared track that permits the skier to experience the exhilaration of gliding over the snow with a smooth motion at a high level of body performance. It is the skiing equivalent of jogging, but without the jarring effect of the foot hitting the ground and with the upper body and arms participating in the exercise. Cross-country skiing may well venture off of prepared tracks, but for our purposes, we consider it to be done for the most part on previously tracked snow with only moderate gains or drops in elevation. Adventure skiing aims to provide more than moderate gains or drops in elevation as the skier enters the world of untracked snow that may vary greatly in its character. Wind and sun work remarkable changes in the consistency of the fallen snow, as does new snow deposited on older snow surfaces. Thus on a downhill run in the untracked snow of the backcountry, the ski may act like an airfoil and glide over the surface of the snow in a flowing fashion in one moment, or it may act like a plow in the next moment, pushing and packing the snow before it as it moves haltingly down the hill.

Backcountry adventure skiers, unless they are supported by a snowcat or helicopter, should carry packs with survival gear and should understand and weigh the consequences of injuries as well as the presence of avalanche hazards. The adventure skier's technique must have a greater degree of authority; there may be little or no room for error; even minor spills may lead to intolerable outcomes. Knowledge of the physics of both snow and skiing should help give skiers a notion of what to expect in the backcountry from the playing field, from the equipment, and from the techniques required to negotiate the challenges of adventure skiing. The chapter concludes by considering briefly under the heading ''The Physics of Survival'' some of the science of weather and of the physical properties one will find in the remote backcountry.

FRICTION: GLIDE AND GRAB

Friction as it is expressed in skiing by its dual attributes, glide and grab, is the subject of Chap. 8. For nordic and adventure skiers to travel effectively over the snow on skis, traction must be turned on and off at will. The ski must slide forward in one instant and then fix itself to the snow surface in the next, while the alternate ski thrusts forward. Our understanding of the ''how and why'' of ski friction and the application of wax or the use of other bottom preparations—whether the goal is to obtain greater or lesser traction—is far from complete, and what we do know about this subject is not widely available. Waxing skis for nordic, alpine, or adventure skiing is both a science and an art. Many a race, both in nordic and in alpine skiing, has been won or lost because a coach or skier picked the wrong wax or applied the right wax incorrectly. Perhaps a little better understanding of what we know about the physics of the playing field and its interaction with the ski—the interface of the snow and the running surface of the ski—may help us cope with this challenging problem.

EPILOGUE: PHYSICS, SKIING, AND THE FUTURE

In Chap. 9 we note that advances in the design and manufacture of ski equipment have changed markedly the way skiing is taught and practiced. The pace of change is so great that skiers who use equipment built for the current season are, in relation to the many more skiers whose equipment is even just a few years old, the skiers of the future. Finally, we consider how recreational skiers may ski so as to avoid injury. We discuss in some detail the physics associated with the several forces that apply to the knee joint in a common fall.

CONCLUSION

In its simplest form, the physics of skiing refers to little more than understanding skiing as the motion of an object sliding down an inclined plane. With that in mind, we invite readers to begin discovering how fascinating the physics of skiing can be. Readers seeking more technical discussions will find what they seek in the technotes. We hope that our readers whose technical interests go not far beyond the motion of that object sliding downhill will find the main chapters of the book—even though they deal with complex, complicated physical properties—relatively accessible and useful for explaining the physical bases of ski equipment and techniques.

In the end, for all of the technical analyses that we offer here, we are well aware that some of the more successful skiing techniques remain, in many ways, somewhat mysterious. But that, after all, is really as it should be when we return to mull over the concept of the triple point, which, while it is a tangible and certain physical property, also has certain mystical quality that allows for a temperature and pressure at which water is at once a solid, a liquid, a gas. Come join us in skiing at the triple point.

REFERENCES

1. C. Fraser makes this conjecture about Hannibal in his book, *Avalanches and Snow Safety* (John Murry, London, 1978), p. 8.

2. For an excellent narrative overview of the history of skis and skiing, see the essay by B. Lash, "The Story of Skiing," in *The Official American Ski Technique* (Cowles, New York, 1970), pp. 3–130. Much of the historical discussion of skiing in the Old World that follows is generally indebted to this source.

3. The definitive, scholarly historical account of skiing in America is E. J. B. Allen's book, *From Skisport to Skiing: One Hundred Years of an American Sport, 1840–1940* (University of Massachusetts Press, Amherst, 1993). See also the shorter account of J. Vaage, "The Norse Started It All," in *The Ski Book*, edited by M. Lund, R. Gillen, and M. Bartlett (Arbor House, New York, 1982), pp. 194–198. The discussion that follows is generally indebted to these two sources.

4. See B. English, *Total Telemarking* (East River, Crested Butte, CO, 1984), p. 29.

5. For more on this and other instances of early American skiing exploits, see the essays by E. Bowen, *The Book of American Skiing* (Bonanza, New York, 1963), Chaps. 2 and 24.

6. J. L. Dyer describes his experiences in his book, *Snow-Shoe Itinerant* (Cranston and Stowe, Cincinnati, OH, 1890; reprinted Father Dyer United Methodist Church, Brekenridge, CO, 1975).

7. See the essay by C. L. Walker, "A Way of Life," in *The Ski Book*, edited by M. Lund, R. Gillen, and M. Bartlett (Arbor House, New York, 1982), pp. 199–205.

8. The term *adventure skiing* to describe skiing in remote areas supported by snowmobile or aircraft was, to our knowledge, coined by Paul Ramer of Boulder, Colorado.

CHAPTER 2

SNOW
The Playing Field

Like other outdoor sports, skiing requires a playing field. That playing field may be provided by nature in the form of a covering of snow. Out of the desire for an adequate playing field, we have contrived to make artificial snow to replace or augment the natural stuff when it refuses to appear in sufficient quantity or, tragically, refuses to appear at all. Most of the alpine skiers of the world practice their sport on carefully prepared slopes on which natural as well as manufactured snow is first packed and then later scored or groomed. Daily maintenance of the slopes' surfaces ensures the uniformly good skiing conditions alpine skiers expect for the price of their lift tickets. Organized nordic ski facilities likewise provide skiers with groomed tracks and courses. When adventure skiers take to the hills or mountains, they likely will seek out particular snow conditions—usually powder, the deeper the better—but they must ski through whatever snow conditions they may find along the way.

Uncertainty about the condition of the playing field is often part of the challenge and fascination of adventure skiing. But in all types of skiing, snow conditions have long provided endless opportunities for much of the small talk among skiers, who may complain, "Ugh, that run was icy!" one day, and then exult, "I exploded through the powder—it was up to here!" the next. To appreciate the complexity of how our skis may skid over icy slopes one day and then glide effortlessly through powdery snow the next, we must look at the physical properties of snow, at how this marvelous substance behaves in its many forms. We must analyze how snow forms in the atmosphere, how it falls to the ground, and how it changes almost immediately upon coming to rest and then continues to change (and is changed deliberately and purposefully when we groom the ski slopes) over time as it lies on the ground as snowpack.

There is no need to carry a long, complicated mathematical expression in our heads as we ski, as the skier does in Fig. 2.1. But skiers who know something about the physical nature of snow—the playing field on which they engage their sport—will understand more about the feel of skiing: why their skis turn in one manner in the early morning and with an entirely different feel, but on the same slope, later that day in the afternoon and why they prefer to ski on snow that the ski report calls packed powder rather than on snow referred to as hard packed. And when adventure skiers head into the backcountry, they will have some better means of judging for themselves the physical nature of the snow cover they ski. They can better choose the preferred waxes for their skis, and they can better make the more

FIGURE 2.1. This skier heads down the hill, his skis lubricated by a film of water that forms under his skis. In his thoughts he mulls over a mathematical formula that we will discuss later in Chap. 8 on snow friction processes. (Colbeck, 1992. Drawn by Marilyn Aber, CRREL.)

critical assessment of the extent of the avalanche hazard that may be present on the naturally snow-covered slopes they ski.

THE FORMATION OF SNOW IN THE ATMOSPHERE

The sequence of events for the creation of snow in nature involves successive condensations of water vapor in the atmosphere. The process begins with the evaporation of water, most commonly from the ocean; or, as is the case in the United States near the Great Lakes, the source of evaporation may be any large body of water or wet land. Once in the atmosphere, cooled water vapor condenses onto ambient particulate matter and returns to its liquid state as water droplets—a process known as nucleation—to form clouds. In the temperate to arctic climatic zones, cloud temperatures are usually well below freezing year around, yet the liquid fog droplets that make up the clouds do not freeze. How then does atmospheric snow form?

Neither the fog droplets that form clouds nor the ice crystals that become snow would form in the atmosphere without the presence of foreign nucleation sites upon which the gaseous molecules of water vapor condense. The world's major source of evaporation of water into the atmosphere, the ocean, also provides the atmosphere with sea-salt particulates that serve as the nucleation sites for most fog droplets. Over land masses, dust or mineral particles may serve that purpose. Once an ice crystal is nucleated and a protosnowflake is created, it will always grow at the expense of the fog, or water vapor, in the surrounding cloud. Water vapor in the clouds is supersaturated relative to an ice crystal, but it is not supersaturated relative to supercooled water droplets. The degree of supersaturation is the ratio of the atmospheric water vapor pressure relative to the water vapor pressure over a surface at the given temperature. Water vapor condenses out of the air and onto the ice crystals—subliming directly from its vapor state to the solid state—rather than increasing the size and number of water droplets in the cloud. And the snowflakes grow thus as they fall. Some high cirrus clouds are composed entirely of minute ice crystals. Evidence of their crystalline nature may be observed when we see "sun dogs" or "halos" around the sun or moon, which are the reflections of light through the ice crystals in the otherwise nearly invisible, extremely-high-altitude, cirrus clouds. These ice crystals do not fall to the ground as snow because they are too small. The air in the high-altitude atmospheric masses in which they form, while it is very cold, is simply too dry for the minute ice crystals to grow into snowflakes that are large enough to fall to the ground.

Precipitation of Snow

The types of snow that grow large enough to fall from the sky as precipitation have been thoroughly catalogued in a variety of ways. In Fig. 2.2, we see an abbreviated classification of the basic forms of atmospheric snow. The topmost panel shows the hexagonal crystal axes of ice; the c axis is the hexagonal symmetry axis, while the a axes point to the corners of the six-sided prism. Notice that snow in the atmosphere may exhibit relatively plain conformations, such as the plate or column shapes shown in the figure. The "star"-shaped, or dendrite, ice crystal, which most of us probably visualize when we think of a snowflake, is only one of the many possible variations on the hexagonal conformation common to all ice crystals.

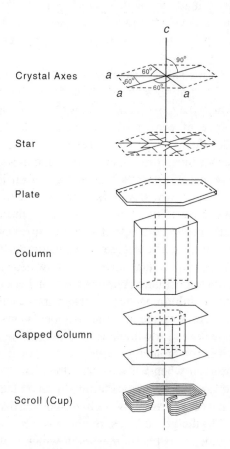

FIGURE 2.2. Principal types of atmospheric snow crystals shown in relation to the crystal axes of ice (Perla and Martinelli, 1979).

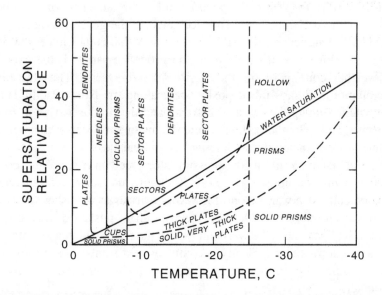

FIGURE 2.3. Atmospheric snow crystal types given as functions of the temperature and supersaturation relative to a flat ice surface at the moment of their formation. (From J. B. Mason, 1971, *The Physics of Clouds*, 2nd ed., Oxford, UK: Clarendon Press. By permission of Oxford University Press.)

In Fig. 2.3 we see the snow types that are probably most familiar to skiers presented in a graph that displays their growth as a function of temperature and supersaturation relative to ice. The conformation of falling snow depends on the temperature and supersaturation at which the snow originates. Thus it is quite easy to determine the passage of a cold front by observing (among other things) changes in the nature of the snow falling to the ground: if platelets or solid prisms of snow change to dendrite flakes, the temperature in the precipitating atmosphere has surely dropped more than 10 degrees. Some 80 types of snow have been catalogued within ten general classifications. As we shall see in the discussions that follow, once snow falls to the ground and accumulates in a pack, it loses its atmospheric conformation in a short time—within a matter of minutes to hours. Heavily wind-driven snow, for example, is composed of completely fragmented flakes that bear little resemblance to their original atmospheric forms.

Atmospheric Airflow

The general atmospheric circulation in the Northern Hemisphere is from East to West in the equatorial region, from West to East at midlatitudes, and again from East to West in the polar latitudes; the same pattern is duplicated in the Southern Hemisphere. This accounts for the general weather pattern over North America: a primarily westerly flow that comes in off the Pacific Ocean and flows over the land mass. In Europe the pattern is the same, with the weather coming in off the North Atlantic Ocean. Thus the highest precipitation and snowfall deposits occur along the North Pacific coast in America and along the coast of Norway and the northern Alps in Europe. Likewise, in the Southern Hemisphere the major snow deposition occurs on the west coasts in southern Chile and New Zealand.

Air circulates around a low-pressure storm cell in a spiraling, counter-clockwise, inward motion; air spirals clockwise and outward around a high-pressure cell. At a warm front, the warm air rises as it pushes the cold air before the front. At a cold front, the cold air is pushed downward much faster as it slides under the warm air before the front, which accounts for the winds and storms that typically accompany cold fronts. These airflows are then modified by the orographic effect of the mountain ranges they encounter.

The flow of storms from off the ocean ensures that the windward sides of continental, West coast mountains have generally large amounts of precipitation, both as rain and snow. Rain falls at the lower elevations, but as the air masses rise to flow over the coastal mountains, the air masses cool and the precipitation changes to snow. If a gas, or air mass, is compressed adiabatically, that is, isolated so no heat can flow in or out, its temperature rises because of the work done in the compression. In a similar fashion, when the air mass expands, it cools. However, if the air mass is saturated with water because it contains fog droplets, on cooling with condensation, the heat of the condensation warms the air. As a result, the lapse rate, or temperature rate change, with altitude for unsaturated air is about 3 °C per 1000 ft; for saturated air, the rate becomes about 2 °C per 1000 ft. Thus on the leeward, warming, or downslope side of a mountain, the air mass will likely not be saturated, so the warming effect is greater than the cooling effect that the same air mass experienced on the upslope side of the mountain.

These features of atmospheric snowfall help determine the locations worldwide of most major ski resorts. They are usually situated in mountains where there is a good flow of maritime, vapor-laden air, and they are usually on the windward sides of the mountains, where the moist, incoming air is rising and cooling, and the depth of deposited snow is greatest. In these maritime climatic zones, the snow has a very different character from the snow that falls farther inland on the higher, colder mountain ranges situated in continental climates. Mountains that receive maritime airflows have heavy snow deposition rates at temperatures close to freezing with high humidity or water vapor in the air. These conditions result in a heavy, dense snowpack that may even have liquid water content, which helps the snow compact and bond. Such snow may easily be formed into a snowball. Mountains in the continental interior receive snow that forms at cooler temperatures in air masses with lower humidity. The result is a lighter, colder snow deposition that features less water content; therefore, the snowpack has little settlement and less internal bonding of the ice grains.

Knowing something about how snow forms as a consequence of particular continental, as well as local, weather patterns should help skiers appreciate how and why there are such variations in the quality of the snow—and thus in the quality of the skiing—available in different regions and even at different ski areas within a region. Skiers might better appreciate the atmospheric origins of some of the nicknames that have been applied to local types of snow, such as "Sierra Cement," "California Concrete," or "Champagne Powder."

MAKING ARTIFICIAL SNOW

No matter where a ski area may be sited, sometimes nature simply does not cooperate by providing adequate natural snow at the proper time and location. Today, major ski areas worldwide make artificial snow as needed for specific runs. As we have seen, natural snow grows in the atmosphere from water vapor condensing as ice, first on nucleation particles, and then in progressively larger amounts on the nucleated ice crystals themselves, until the flakes are massive enough to fall as snow. Artificial snow mimics some parts of this process by using high-pressure air guns to inject liquid water 20–30 ft into the air over a ski slope, creating an artificial fog of minute water droplets. This fog of liquid water droplets is encouraged to freeze by mixing a nucleating agent—most often a harmless bacterial protein—with the water. The bacterial protein offers an especially effective nucleation site that helps initiate from inside the water droplets the conversion of the supercooled fog directly into ice crystals at the relatively warm temperature of about −2 °C (28 °F). As soon as it is cold enough, today's ski area operators can make a playing field of snow without a single cloud in sight. An added benefit is that artificial snow makes a superior base layer for heavily used ski runs because it is much denser than natural snow. Artificial snowflakes have the conformation of rounded ice grains, and they pack into a strong mass, much like a hard snowball.

The physical and engineering requirements for making artificial snow have been known for some time [1]. The cold water that emerges from the

high-pressure nozzle of a "snow gun" appears as a cloud of minute droplets having diameters regulated by the gun, usually in the range of 100–700
microns (μm). These droplets are shot some 20–30 ft into the air. One
hundred microns is about 0.004 of an inch, so the fog is composed of a very
fine mist. Just as it does naturally in the atmosphere when a wet air mass
slams into the windward side of a mountain range, the air and water mixture
cools by adiabatic expansion when it exits the nozzle; however, the crucial
cooling needed to freeze the water droplets into artificial snow must come
from the surrounding atmospheric air.

All natural water has foreign particles in it that may serve as nucleating
agents, but when water containing only natural nucleating agents is used to
make artificial snow, its freezing temperature can be relatively low, aver-
aging about −8 °C or 18 °F. In Fig. 2.4, taken from the marketing literature
of a commercial supplier of a bacterial nucleation product, we see data that
report the varying freezing points of water samples taken from several ski
areas compared to the freezing points for the same water samples when they
are mixed with a bacterial protein. Adding the bacterial nucleating agent to

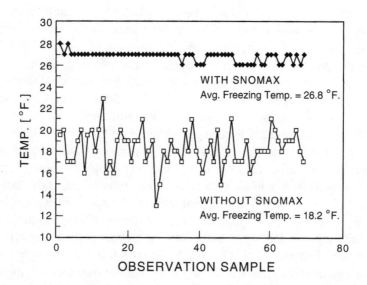

FIGURE 2.4. The freezing temperature of more than 60 water samples taken from
natural sources. The line above shows that adding ice-nucleating protein to the water
greatly regularizes the freezing temperature and yields the relatively high average
freezing temperature of 26.9 °F. Natural water samples have 0–115 nucleation sites per
milliliter; the nucleating protein raises that number to about 250 000 nucleation sites per
milliliter. (Reprinted with permission from marketing literature of Snowmax Technologies,
Rochester, NY.)

the snowmaking mix raises the freezing temperature of the water samples to 26.8 °F or −2.9 °C, which makes it easier to make artificial snow.

When water droplets of would-be artificial snow are shot into the air, they must freeze before they fall to the ground or evaporate; this freezing can only be accomplished by the temperature of the surrounding air. If the water droplets fall to the ground still in their liquid state, they freeze immediately upon contact with the ice grains present on the ground to form a solid slab of ice—not the ideal playing surface for skiing. The time it takes the seeded droplets to cool and freeze must be less than their hang time in the air. Droplets in the range of 100–700 μm when shot from a snow gun will hang in the air for about 15 seconds, so the presence of the bacterial nucleating agent helps ensure that the droplets will freeze in the air, even at temperatures no more than a few degrees below freezing.

Note how artificial snowmaking differs from the formation of atmospheric snow. In the clouds, the hang time of the nucleated, protosnowflakes is large. The grains of atmospheric snow grow by vapor accretion, depending on the relative humidity or the degree of supersaturation of the water vapor in the air relative to the snow grain. The fog droplets that form artificial snow must freeze quickly, and their conformation is not augmented by the humidity of the air around them. This makes artificial snow especially appropriate for the drier, colder atmospheric conditions typically found at ski areas sited in the mountains of the continental interior.

It has been observed for some time in nature that biological structures may provide a lattice-matching template for the formation of ice crystal lattices. Lichens may provide such material, but the most frequently observed material comes from several bacterial organisms [2]. The bacterium strain *Pseudomonas syringae*, originally isolated from a corn plant, is widely distributed in the environment and makes a good ice nucleating agent. These bacteria are cultivated and then freeze-dried so that the remaining protein mass contains no live cells. The protein present in the cell wall membranes of these organisms is the active nucleating agent, and it varies in composition. Figure 2.5 shows how the bacterial protein is sited on the cell membrane. The protein is composed of amino acid chains of eight units that couple to form 16-member chains, of which three chains are coupled to a 48-unit structure. These 48-unit structures are each composed of 6 of the 8-unit building blocks of protein, which coil to form a hexagonal array that repeats with increasing fidelity toward one end. The high fidelity of the *P. syringae* protein chains, coupled with their helical protein structure, makes them good templates for the formation of ice crystals.

The exact mechanism operating at these template sites that facilitates the alignment of water molecules to form ice crystals is not well known. The

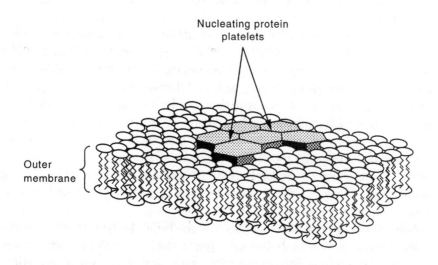

FIGURE 2.5. The structure of a bacterial membrane with a cluster of hexagonally arrayed protein platelets on the surface. The cluster of platelets serves as the nucleation site for the initiation of ice crystal growth. (Modified from Warren and Wolber, 1991. Copyright Blackwell Science Ltd., Oxford, UK.)

bond length between the carbon or nitrogen atoms in the bacterial protein is comparable to the bond length between the oxygen atoms in ice (H_2O), making these protein structures essentially rather large analogs of ice crystal lattices. Molecular bonds that hold the water molecules in an appropriate geometric array must be present so that further accretion of water vapor may occur, permitting the embryonic ice crystal nucleates to grow quickly into full ice crystallites to which other water molecules accrete directly.

Various studies of artificial snowmaking demonstrate that the liquid water content in artificial snow is small, and that the density of artificial snow produced with the use of bacterial nucleates ranges from 400 to 440 $kg\,m^{-3}$. The density of newly fallen natural snow is seldom above 100 $kg\,m^{-3}$, one-fourth the density of artificial snow. As we shall see, it takes varying amounts of time, depending on the conditions present, for natural snowflakes to change into the rounded ice grains that will make the dense snowpack needed for a strong skiing base. Artificial snow has both high density and uniformly rounded grains from the start. In a very short time it can be spread and groomed over an alpine ski run to make a base for skiing that exhibits a strength much superior to natural snow.

THE SNOW COVER ON THE GROUND

The falling snow is distributed on the ground by the wind and by the nature of the local topography and vegetation cover. As we have seen, snow deposits over a mountain range will be, in general, larger on the windward side of the mountains and significantly smaller on the leeward side, diminishing with distance until the snow-bearing clouds encounter another range and the windward–leeward process begins again. Just the opposite may occur over certain ridges and valleys. The higher velocity winds on the windward side of a ridge may suspend the falling snow in the air and scour the ground-cover snow, whirling it into the air if the snow cover is not sheltered by vegetation. When the wind velocity diminishes on the leeward side of the ridge, the augmented volume of suspended snow settles out in a massive cushion. If one knows the general weather pattern for an area, the topography, and the vegetation, it is quite easy to identify major snow accumulation areas. Backcountry skiers should realize, for reasons to be discussed below, that these areas of high snow accumulation are also areas with a high avalanche hazard. Unfortunately, they are also the prime, powder skiing slopes, greatly preferred over the windward slopes whose surface-eroded snow will be hard packed.

Although the snow falling from the sky always occurs as six-sided plates, prisms, needles, columns, or dendritic, six-armed lattices, these forms never remain for very long once the snow is on the ground. Because the solid ice in the snowflake is relatively close to the melting point, or, more precisely, the triple point, the convex points and edges of the snowflakes tend to evaporate. Thus the crystals of newly fallen snow lose their sharp corners and edges and become rounded grains of roughly uniform size, bearing almost no resemblance to the crystal conformations of atmospheric snow that were illustrated above in Fig. 2.2. This process—known as equitemperature metamorphism—may be seen in Fig. 2.6, which shows a sequence of microphotographs of newly fallen snow. The sharp points, edges, and corners of the young snow on the left disappear rapidly. The concavities of the dendritic crystals fill in; the needlelike forms become much more rounded.

As the snow changes its form and settles, its bonding strength may be reduced initially because the rounded grains of snow will roll or slide over one another. On a steep slope, this process creates an unstable snowpack, the key to avalanche hazard. Until newly fallen snow is compacted, either under its own weight or as a result of continued metamorphism, it can be unstable. When the mass of snow is compacted, the grains contact one another and ice bridges grow. The grains of snow become sintered, or

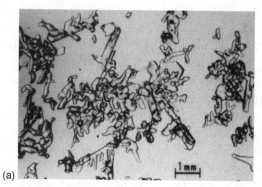

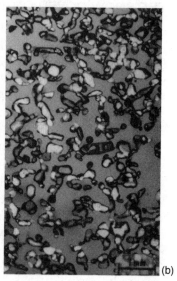

FIGURE 2.6. Equitemperature (ET) metamorphism of newly fallen snow. The faceted, dendritic arms of the new snow grains in the left frame have disappeared in the right frame, metamorphosing into rounded, sintered grains. (Reprinted with permission from Colbeck *et al.,* 1990. Photographs courtesy of E. Akitaya.)

welded, together. This is the process that occurs when we make a snowball with moderately dry snow or pack soft snow under our skis or boots. Thus in the early season before the steeper ski runs at an area have adequate snow cover, maintenance crews go up the hills and pack the snow of the steepest slopes either by ski or by boot. The packing of the shallow snow on the steep slopes improves the strength of the snow pack; it also helps new snow bond to the existing snow pack.

In nature, the snow cover consists of many different layers, representing separate snowfalls and the varying effects of the weather between storms. Newly fallen snow has a density of $50-100$ kg m^{-3}, compared to the density of solid ice, which is 920 kg m^{-3}. The density of older, naturally packed snow ranges from that of new snow to a value equal to about 3/10 the density of ice, or about 300 kg m^{-3}. Natural snow that has been ground and packed at a ski area and artificial snow have densities ranging up to about 500 kg m^{-3}, at most.

Whether in nature or at the ski area, the snowpack is mostly air, very porous, and saturated with water vapor. The air pores in a snowpack do not begin to close significantly or become isolated until the snow has been so compacted over such a long period of time that it becomes glacial ice. All glacial ice was once newly fallen snow, and it retains air locked within it, either in tiny, isolated bubbles or entirely dissolved in the ice itself. The density of glacial ice is around 800 kg m^{-3}. Basically, snow cover on the ground exhibits one or more of three structures: ice grains of comparable size, depending on their age, accreted together in clusters, the result of rounded-grain metamorphism; larger, separated ice grains that are loosely bound, the result of faceted-grain metamorphism; and, finally, clusters of ice grains into which meltwater has percolated and refrozen, the result of melt–freeze metamorphism.

Rounded-Grain Metamorphism

Rounded-grain metamorphism has also been called equitemperature or ET metamorphism because it occurs when the overall temperature change in a snow pack is small. Rounded-grain metamorphism changes the snow crystal structures in two significant ways: the snow grains become significantly rounded, and bonding occurs at the contact points between the grains, so the snow mass becomes progressively stronger over time. Whether the snow grains are packed by the weight of the overburden of snow or by some artificial means, the snow formed by this process (sometimes called ET snow) has the structure shown in Fig. 2.7.

With compaction, the touching ice grains form composite structures with concave pockets and convex points that, taken together, describe an interface between the solid ice and the gaseous water vapor occupying the pockets between the ice grains. The curvatures of these ice–vapor interfaces dictate the vapor pressure differences indicated in Fig. 2.7 by the + and (−) symbols. That is, at the convexities—the isolated points of the snow grains that do not touch each other—the vapor pressure is above the average; at the concavities—the pockets formed by contact between the grains—the vapor pressure is below the average. Even though on the average and taken at the macro level, the temperature throughout the snow pack is fairly uniform; on the micro level at which the individual snow grains pack together, the convexities must become cooler because of the heat needed for vaporization. And at the concavities where the vapor freezes, the heat of condensation is released to warm the ice.

Ice is a relatively good conductor, so heat flows in the solid ice from the relatively warm concavities to the relatively cool convex regions to maintain the process. When the water mass is carried as vapor, heat is also transported; that heat must, by conduction, flow back to the source to maintain the process. Thus water mass is carried from the higher-pressure, convex points to the lower-pressure, concave regions continuously by vapor diffusion, which depends on the vapor pressure difference, the Kelvin (or absolute) temperature, and the distance. The vapor-pressure differences as well as the diffusion rates depend on the temperature, so the process slows when the temperature drops. Likewise, as the average distance between

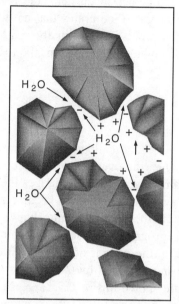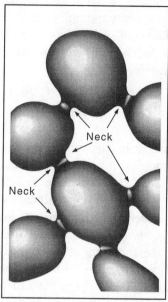

FIGURE 2.7. A distribution of ice grains in an equitemperature snowpack with the overburden pressing the grains into contact. The + symbols indicate where the vapor pressure is high because of positive curvature, and the − symbols show the regions that have low vapor pressure because of the negative curvature. The vapor migration and the growth of sintering ice necks are illustrated in the adjacent frame (Perla and Martinelli, 1979).

convexities and concavities decreases with compaction of the snow, the diffusion rate increases. At −40 °C the process almost ceases; near 0 °C, it is quite fast. Fresh snow near 0 °C will harden noticeably in a matter of seconds when it is compacted under a ski or boot.

The tendency for water vapor to migrate to the contact regions between snow grains to form ice bonds is called *sintering*—the snow grains are welded together naturally by the process of equitemperature metamorphism. We observe the same process in our freezers when the older ice cubes in the storage bin seem to melt and bond together, even though the temperature in the freezer compartment stays relatively constant. So too, in nature, larger snow grains tend to grow over time at the expense of smaller ones, creating a harder, more cohesive snow pack as the snow is compacted, either by the pressure of the overburden or by the operation of mechanical compaction equipment. When ski areas groom their slopes by grinding and rolling the surface of the snow, they create the conditions needed for the snow to harden into a dense, well-bonded mass that will not become gouged or rutted by the many skiers who use it. In the morning after the grooming crews have raked and rolled the slopes, the snow surface is relatively hard, and our skis may chatter as we edge them for a turn; only after the snow becomes abraded by hours of ski traffic does the surface become covered with a layer of relatively loose snow that our skis can carve more smoothly.

Faceted-Grain Metamorphism

The second major type of metamorphism in the ground-cover snow is driven by a temperature gradient in the snowpack itself, which leads to the creation of snow crystals that exhibit faceted grains. Such snow is also called TG (for temperature gradient) snow. The temperature gradient usually exists near the ground, which acts as a heat source (see Fig. 2.8).

The snowpack is bounded on the top by the atmosphere and on the bottom by the ground cover. At the snow surface, heat from the sun and atmosphere is absorbed or scattered, but heat is also radiated out of the snow. At the bottom of the pack, the ground heat, except in arctic regions, raises the temperature to 0 °C when the snow provides the necessary insulation from the cold atmosphere. As a general rule, 6 in. of snow cover provides sufficient insulation for such warming.

Because of the more moderate climate and the greater compaction of the

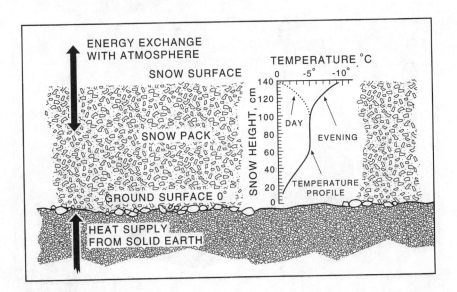

FIGURE 2.8. Temperature profile of a snowpack. Temperature gradients exist at the top and at the bottom, near the surface of the snowpack and at the ground surface. In the middle, the snowpack is at an equitemperature state (Perla and Martinelli, 1979).

water-laden snowfall, mountains in maritime climates have snowpacks with relatively small temperature gradients, and TG snow is not often seen. Temperature-gradient snow is most prevalent in areas of continental climate, where the air temperatures are low and the snow deposits are not too thick. Under these conditions, the temperature gradient exceeds a limiting value and the snow is not subject to overburden compaction or to meltwater destabilization.

Sinter-bonded snow at the bottom of the snowpack is warmer than the snow above, so the vapor pressure is larger there. Water vapor is driven across the voids from the bottom to the top of the pack by the temperature gradient and by the concentration gradient in the process illustrated in Fig. 2.9. Rising water vapor condenses on the snow grains above, transforming them into the faceted grains associated with TG snow. These snow grains have prismatic angles and surfaces and exhibit the characteristic, sixfold symmetry of ice.

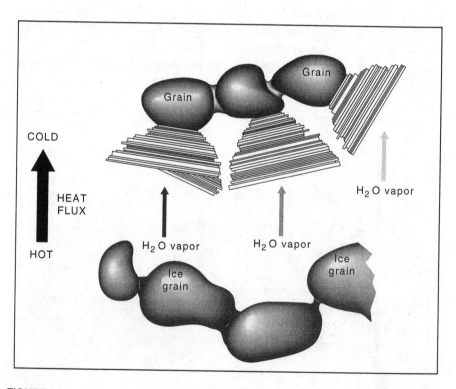

FIGURE 2.9. The transport by convection of water vapor from the warm, sintered grains below to the colder grains above. Condensation forms the prismatic laminar facets (Perla and Martinelli, 1979).

Faceted-grain metamorphism is a relatively slow process. It will go on for months if the snow mass is undisturbed by compacting forces. Because the source of heat, the ground, and the heat sink, the atmosphere, are constantly maintained, heat does not need to be conducted back into the snowpack to keep the process going. The microphotographs in Fig. 2.10 show faceted grains of TG snow growing in this manner.

The characteristics of faceted-grain, TG snow are quite different from the characteristics of rounded-grain, ET snow. While the density of the snow does not change, the individual snow grains grow in size and the bonds between the grains per unit volume decrease. Thus the cohesive strength of the snowpack decreases markedly. The snowpack may become so unstable that it cannot support the weight of the overburden snow or the weight of even a single skier. Once a snowpack collapses, the TG snow has no cohesion, and it runs like a fluid. TG snow behaves much like very dry sugar; it is likely the ball bearings upon which most slab avalanches run.

Perhaps the essential element in evaluating the avalanche hazard of a given snowpack is to observe the layers of temperature-gradient snow in relation to the layers of equitemperature snow. The ET and TG processes usually occur so that, in general, ET snow is found in the upper layers of the snowpack and TG snow in the lower layers. The ET process is quite rapid, and the TG process is fairly slow; ET snow makes for a strong snowpack, and TG snow makes for a weak snowpack. The two types of snow are nearly opposite in their characteristics, yet by the nature of their formation, they are found together, one on top of the other, in the deep snowpacks of continental mountains.

In general, a strong, thick layer of ET snow provides a safe, reliably bonded playing field for adventure skiing in the backcountry. Problems are

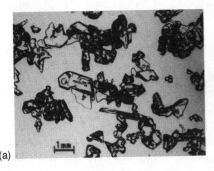

(a) (b)

FIGURE 2.10. Temperature-gradient (TG) snow in successive stages of growth. Notice the change in scale from (a) to (b). (Reprinted with permission from Colbeck *et al.*, 1990. Photographs courtesy of K. Izumi.)

caused by the layers of TG snow, and predicting how these layers of TG snow will behave remains fraught with uncertainty. TG snow may form in thin layers near the top of a snow pack when new, low-density snow falls on an old snow deposit and the air temperature drops, or when the slope is especially cold, facing away from the sun to the North. In this instance, the old snow mass serves as the heat source, and the layer of new, low-density snow is thin, so the temperature gradient is high. Under these conditions, TG snow forms rapidly and then becomes overlain with new snow deposits. The thin layer of TG snow will always be weak, and it easily serves as a glide plane for the snow deposits above. Thus a two-stage avalanche may occur: the top layers run first with the lower layers giving away later.

In backcountry skiing, it is essentially impossible to negotiate a slope that has a deep layer of TG snow. If the overburden of ET snow breaks and the skier sinks through into the TG snow, the skis will plow under the snow and will not rise back to the surface. The TG snow does not compact, so turning maneuvers cannot be executed because the ski cannot compact the snow. Finally, the resistance to forward motion from the weight of the overburden snow can be so great as to prevent any motion, except on the steepest of slopes. Such was the case in the winter of 1991–92 in Colorado. In November and December large snowfalls were followed by six weeks of cold, dry weather, followed by about two weeks of modest snowfall. A snow pit dug at 11 100 ft elevation on the flat had 3 ft of faceted-grain, TG snow with an overburden of $1\frac{1}{2}$ ft of low-density, rounded-grain, ET snow. Skiers making a track in this snow fell through the relatively thin layer of ET snow and sank about 2 ft before compacting the snow into a track with enough stability to ski. Trying to ski anything but a packed track was nearly impossible; for the skier at the front of the ski tour party, the skiing on this snowpack was more like snowshoeing. High avalanche hazard existed, and the hazard increased as the winter wore on and more snow was deposited.

Melt–Freeze Metamorphism

Liquid or free water in the snowpack causes rapid changes in the snowpack's structure and strength. The free water may come from rain falling on the surface of the pack, or it may be meltwater generated by the radiated heat of the sun shining down on the snow's surface. Rainwater will not melt much snow; unless the rain comes from an extraordinarily warm weather system, the available heat is usually just too low. If the snowpack itself is fairly cold, rain falling on it will warm the snow somewhat as the rain's relative warmth diffuses into the snowpack and the free rainwater either freezes on the surface of the pack or percolates into it and freezes under the surface. Solar radiation is the chief source of heat for the generation of free water in a snowpack, particularly in continental snow masses. Snow is an extremely good scatterer of sunlight when it is clean, so little radiation is absorbed. If, however, any free water is present in the snowpack, the water will absorb considerable amounts of radiation, and, with the resulting rise in temperature, melting hastens.

The movement of free water into the snowpack is enhanced by the shapes of the ice grains themselves. Just as curvature affects the vapor pressure at the solid–vapor interface, so it also affects the freezing point of the liquid–solid interface. When the pore space in the snow is filled with liquid so that only liquid–solid interfaces exist and the pressure is constant, the melting temperature of a snow grain depends on its radius: because there is a surface tension between the solid and liquid states, the smaller the snow grain, the lower its melting temperature. Small grains have lower temperatures than the larger grains, so heat flows to them, generating meltwater, which freezes onto the larger grains; the heat of fusion that results from the phase change from liquid water to ice is conducted away. The ice bonds that connect the individual snow grains in the pack have a curvature opposite to that of the grains. They also have a higher melting temperature, but they do melt.

The process of melt–freeze metamorphism is far faster than either rounded-grain or faceted-grain metamorphism because mass does not have to be transported; there is plenty of free water that flows where it will. The liquid also enhances the heat transport necessary to sustain the process. The larger ice grains literally gobble up the smaller grains. The process of melt–freeze metamorphism takes a matter of hours, not days. Water spreads out in the crust layers of the snow and may run laterally rather than percolate down into the pack. In the two photographs in Fig. 2.11, we observe how bonded grains of melt–freeze metamorphism in a pack (a) become rounded, free of ice bonds, and lubricated by meltwater. Everything is ready to move. There is little or no strength in such a snowpack, so the mass may avalanche. If, when night falls, the temperature goes down enough so that the free water refreezes, the snow mass becomes extremely strong and hard, exhibiting the characteristics more of ice than of snow. New snow deposited on such a frozen surface does not bond at all, so locating ice crusts helps the ski patrol or snow avalanche expert find the slide planes needed for the controlled release of avalanches.

Melt–freeze metamorphism is also used to condition snow for specific skiing needs. Late in the day, a ski race track may be sprinkled with salt or one of several other chemical compounds that release heat as well as depress the freezing point when they go into solution with water. A liquid-

(a) (b)

FIGURE 2.11. Ice grains subject to meltwater metamorphism and refreezing. (Reprinted with permission from Colbeck *et al.*, 1990. Photographs courtesy of S. Colbeck.)

water solution forms, which enhances the melt–freeze metamorphism, and then the subsequent nighttime freezing makes the snow surface very hard and compacted for the race.

Recognizing the different structures that snow grains may exhibit and understanding their relative strengths are especially important for avalanche forecasting. The adventure skier who skis the backcountry must be familiar with avalanche hazards and be able to read the signs that nature provides. With a small hand magnifier, an adventure skier can gain a good feel for the snow structure and its stability by digging a pit into the snow for several feet and then observing the types of snow layered in the pack. Even on a groomed ski slope, it can be an interesting exercise to look at some new snowflakes that stay frozen long enough to be observed on a woolen glove or scarf and then compare the snow deposited in the pack at the different depths exposed on the edge of a run. Look too at the mirrorlike surface of a ski track created by the passage of a ski. You should recognize the variety of snow grains created by the metamorphic processes discussed above [3].

THE WATER MOLECULE

All of the processes described above have their origins in the properties that derive from the atomic structure of the water molecule and from the molecular structure of the liquid, solid, and vapor states of this fascinating substance. The atomic composition of the water molecule is H_2O. The hydrogen atom, H, has one electron orbiting its positively charged nucleus, or proton. The oxygen atom, O, has eight orbiting electrons because its nucleus contains eight protons along with eight neutrons, giving it eight positive charges that must be balanced by the negatively charged electrons. Two of the electrons compose an inside shell and do not react with the electrons of other atoms to form the chemical bonds required to create molecules. All of the remaining six electrons in oxygen may participate in chemical bonding, and they occupy two shells: one contains two electrons; the other contains four electrons.

In the water molecule, eight electrons form chemical bonds: six from the single oxygen atom and two from the two hydrogen atoms. Two electrons from the oxygen atom pair with the electrons from the two hydrogen atoms to form covalent bonds between the hydrogen atoms and the oxygen atom. The two hydrogen atoms bond in such a manner that the angle between the bonds has an arc of 105°; thus the hydrogen atoms are arrayed on one side of the oxygen atom, giving the molecule its characteristic, asymmetrical structure. The remaining four oxygen electrons couple in pairs that lie on opposite sides of the oxygen nucleus, perpendicular to the plane of the molecule as shown in Fig. 2.12, a qualitative representation of the atomic structure of a single water molecule. The dimensions given are in nanometers (nm); 1 nm is a billionth of a meter.

All of the stable building blocks of nature—electrons, protons, and neutrons—have angular momentum; that is, they spin like tops around a central axis. Angular momentum is an intrinsic, quantized property, like particle mass. Just as magnets will attract or repel each other depending upon the relative orientation of their charged poles, atomic particles have preferential orientations relative to each other when they come together as pairs. In Fig. 2.12, the symbol ↑↓ denotes the relative angular momentum of the pairs of bonded electrons in the water molecule. Note that this illustration is a heuristic, qualitative representation; that is, the electrons do not actually follow the circumscribed orbital paths depicted in the illus-

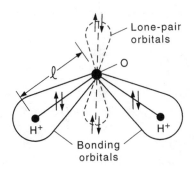

FIGURE 2.12. Atomic structure of the free water molecule: one atom of oxygen (O) bonds with two atoms of hydrogen (H). The distance between the atoms is given as ℓ and equals 0.096 nm. The bonding orbitals and the lone-pair orbitals form a roughly tetragonal system.

tration; they are in fact smeared out in the space between the nuclei of the molecularly bonded atoms.

In the liquid state, if another H_2O molecule were to happen by our single, free molecule of water, a pair of oxygen electrons in the unbonded orbit of one of the molecules would be attracted to the positive charge of one of the hydrogen nuclei in the other molecule, and the two molecules would come together to form a pair, called a *dimer* (pronounced *die-mer*) (see Fig. 2.13).

When these dimers meet, they form bonded strings and clusters of water molecules called *polymers*, creating polymerized water. One water molecule has a total of eight electrons, six from the oxygen atoms and two from the hydrogen atoms, which may pair to make four covalent bonds around each oxygen atom. In completely polymerized water, every oxygen atom has four bonds with surrounding hydrogen atoms, and through those bonds the atom bonds to four other oxygen atoms. These electron-bond pairs electrostatically repel one another, so the bonds form a tetrahedral structure. Molecules of such completely polymerized water would be arrayed in tetrahedral iterations that would form large polymer strings (see Fig. 2.14).

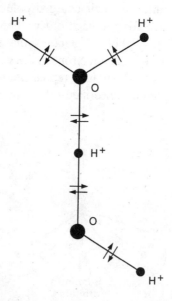

FIGURE 2.13. Schematic representation of the configuration of a twisted dimer of water molecules. The positive hydrogen atoms (H^+) repel one another, so the bottom molecule is twisted off to one side relative to the upper molecule. Additional water molecules may attach to the free (H^+) atom through their lone-pair orbitals to form polymer strings of water molecules.

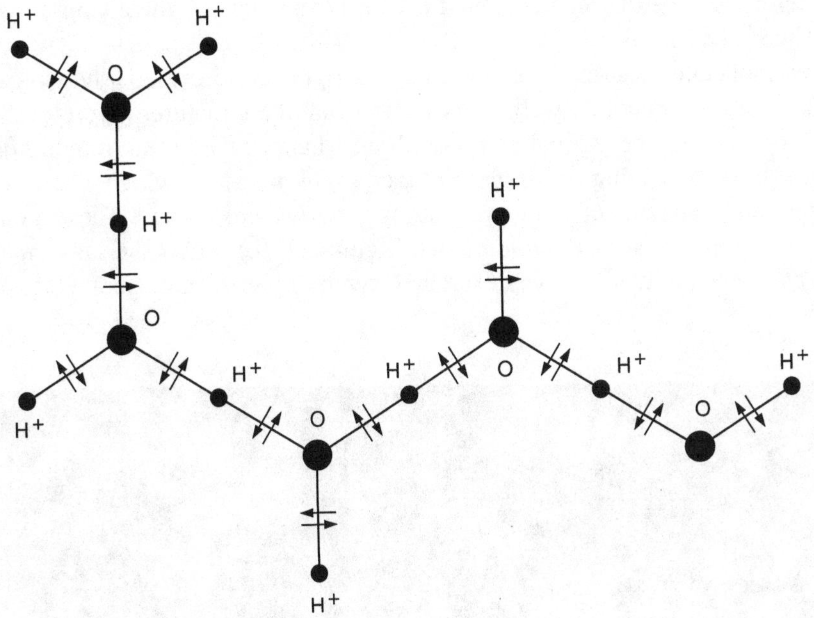

FIGURE 2.14. Chain of polymerized water.

Water molecules form the playing field for skiing, and at the temperatures and pressures at which we ski on them, we encounter water molecules in all three of their phases or states: as a vapor, as a liquid, and as a solid. Each of these states exhibit different structures and different properties. In the vapor phase, the water molecules are widely separated, so much so that they exist as single molecules. Our interest in the vapor state of water will focus mostly on how the molecules move from the vapor state to either the liquid or solid states. Let us examine the physical processes associated with water that take place at the atomic and molecular scales, starting with the solid state.

The Solid State

If we arrange tetrahedra in a three-dimensional lattice, we quickly find that if we place two tetrahedra together, point to point, and then place three such pairs in a ring, we create a structure of interlocking hexagons that closes upon itself in a lattice with a minimum of distortion of the bonds.

Figure 2.15 shows a model of the interlocking, hexagonal structure of a microcrystallite of ice, designated I_h, for hexagonal ice, the common form of ice found in nature.

The model illustrates part of a microscopic I_h ice crystal, showing the faces of the crystal as well as its internal, lattice structure. Figure 2.15(a) looks down the hexagonal array, or along what is called the symmetry or c axis. The connecting rods represent the bond lines along which a hydrogen atom and two pairs of covalent bonding electrons are located. These atomic features are not shown in the model. Figure 2.15(b) shows the same model seen perpendicular to the hexagonal symmetry axis and parallel to the

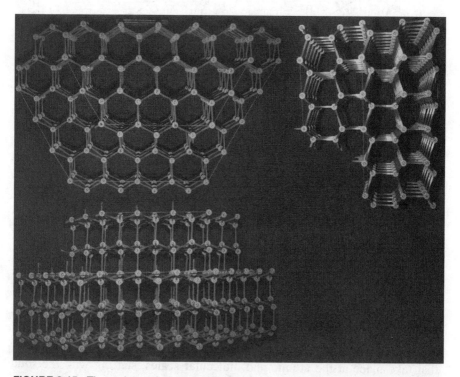

FIGURE 2.15. Three views of a lattice model for a microcrystallite of ice I_h showing only the oxygen atoms. The connecting rods between the nuclei represent the bond lines upon which a hydrogen atom and two pairs of covalent bonding electrons, which are not shown, are located. Only half of the hexagonal crystal is shown; the step is a characteristic of the prism faces. View (a) is along the c or hexagonal symmetry axis. View (b) is parallel to a prism face, or along an a axis perpendicular to the c axis. View (c) is a front elevation seen perpendicular to a prism face. The distance across the faces of each unit hexagon in the lattice is 0.452 nm or 10^{-9} m. The basal planes are normal to the c axis. A selection of surface bonding sites for water is shown by the short stub connectors. (Model fabricated with Zometool structural system, Zometool, P.O. Box 7053, Boulder, Colorado 80304, electronic mail: Zometool@aol.com.)

prismatic faces of the structure. Figure 2.15(c) is an elevation view normal to a prism face. The dimensions of the lattice are 0.452 nm units between the faces of the hexagons, with 0.734-nm separation from one plane to the second plane above, because the lattice repetition interval is two horizontal planes, as seen in the figure. Recall from Fig. 2.13 that the size of a free water molecule is only about 0.2 nm across, so free water molecules can readily diffuse through the crystal structure of ice.

In Chap. 8 we will discuss the organic, carbon based molecules—for the most part waxes—used as ski bases, so let us consider how corresponding bonds for carbon form. Carbon has a total of six electrons, but two electrons in the inner shell are not chemically active, so each atom has four active electrons, available to participate in bonding. Thus carbon atoms can bond to other carbon atoms when each provides one electron. With four bonds possible, carbon may form the same tetrahedral configuration that we observed around the oxygen atoms in ice. Graphite is a form of bonded carbon atoms that exhibits the same hexagonal structure as ice. In both graphite and ice, a small rearrangement of the bonds generates a cubic symmetry. Thus diamond is the cubic version of carbon, and ice I_c is a cubic version of ice, found only in the high stratosphere. Cubic ice changes to ice I_h—the hexagonal version—when it is warmed. Both graphite and ice shear readily along the axis perpendicular to their hexagonal structures; graphite serves as a good lubricant because of this property. It may also be true that the sliding of skis on ice is facilitated by the same shear failure property; certainly in glacier flow something like this may occur because large ice crystals form with their hexagonal axes all aligned in one preferential direction.

The Liquid State

As we have seen from the discussions above, as water vapor cools, the water molecules may condense on the liquid surfaces of existing water droplets or onto foreign particles to form fog. When the random thermal motion of these molecules of liquid water becomes subdued at still lower temperatures, the water molecules will coalesce to form long strings or chains of water molecules called polymers. The snaky, string-shaped molecules of water do not fit together particularly well, but they become more compatible as their individual motions subside with cooling. Thus the density of liquid water increases as the packing of the molecules improves; polymerization increases as the temperature falls, and longer strings of water molecules form. So far this process is similar to the behavior of all liquids. At 10 °C these polymer chains may contain on average as many as ten water molecules.

Then something unusual happens. On cooling below 4 °C, the large polymer strings of liquid water molecules can no longer pack together, and the density of the water actually decreases as the temperature falls to the freezing point. Thus a lake cannot freeze over until all of the water in the lake reaches at least 4 °C; then the colder liquid, being less dense, floats to the surface where it can cool further to the freezing point. Colloquially, we say that the lake "turns over," and that is more or less what happens. At the freezing point the long polymers of water molecules strung out in their liquid state must rearrange into the hexagonal lattice structure associated with ice. That reconfiguration involves forming a structure that requires more space than is needed for the same number of molecules in their liquid state. Thus the density of water drops from 1000 kg m^{-3} at 4 °C to 999.84 kg m^{-3} at 0 °C. The ice that forms at 0 °C has a density of 916.7 kg m^{-3}. Solid ice floats in water, with about 8% of the volume of the ice rising above the water. Water is one of a very few substances known for which the solid phase is less dense than the liquid.

It is also significant that a large amount of heat must be removed from the liquid state to accommodate the structure change to the solid state. The kinetic energy of random motion that is present in the liquid state is not present in the solid state. Also, potential energy is reduced with the bond formation that accompanies the phase change from liquid polymers to crystal lattices of ice. If the necessary heat is not removed, or for some reason the water molecules cannot align themselves, water can stay in its liquid form well below 0 °C. In fact, most of the clouds in the atmosphere are composed of liquid water drops with temperatures that go down to at least −20 °C.

The Phase Diagram

To understand more fully the physical processes involved in the formation of snow and in the deposition and metamorphism of the ground snowpack, we must consider the pressure and temperature relations associated with the phase changes of water as it moves through its three states: vapor, liquid, and solid. Pressure and temperature determine the phase state of water.

Consider a container in the form of a cylinder fitted with a plunger. The container is filled with a given mass of water. There is no other "air" in the container, so liquid water and water vapor occupy the whole volume (see Fig. 2.16). Observe what happens when we heat or cool the system while

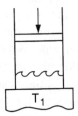

FIGURE 2.16. A cylinder containing liquid water and water vapor in contact with a heat reservoir at a temperature T_1. The cylinder is fitted with a piston that may be raised or lowered to affect the pressure in the system.

maintaining the pressure at a constant value. When we add heat so that all of the liquid water vaporizes, the piston must move upward in the cylinder to keep the pressure in the container constant. With constant pressure in the cylinder, the volume of the water (now entirely in its gaseous, vapor state) depends on the initial amount of water and the temperature to which we heat it.

Now, with only vapor in the cylinder, we will remove heat from the system while once again we hold the pressure constant by moving the piston downward in the cylinder. At some temperature, called the dew point, the vaporous water will begin to condense into a liquid. As the system is further cooled, more liquid water will condense out of the vapor, but the temperature in the system cannot change because of the heat that is released from the vapor as it condenses. Only when all of the water has condensed into a liquid by the removal of heat from the system can the temperature of the liquid water itself be further reduced. Now the cylinder containing only liquid may be cooled until the liquid begins to freeze. Again, with the phase change from liquid to solid, some heat is released and must be removed (that is, the system must be cooled). When all of the liquid is frozen and the released heat is removed, then the temperature of the solid may fall further. The volume of the system may change, but the mass of water is constant.

Throughout this process the piston moves upward and downward in the cylinder to maintain a constant pressure during the phase changes. First, the piston must move upward as the water mass in the system vaporizes entirely and the gas expands. Then the piston moves back downward into the cylinder because the volume of any amount of condensed liquid water is much smaller than the volume of the same water mass in its vapor phase. Once only liquid is present, the system may be cooled with only a small decrease in volume to maintain constant pressure. As the liquid water cools, the piston barely moves, but it continues to move downward. When the phase boundary between liquid and solid is crossed and ice starts to form, the piston reverses its direction. Because the density of ice is lower than the density of liquid water, the piston must again move upward in the cylinder—be withdrawn—to allow the frozen water to expand until all of the liquid water is frozen. Many a bottle of water or automobile engine has been shattered or cracked by this effect.

So far we have considered this system in the presence of a constant pressure. More interesting is the effect of pressure changes on the phase-change process. Figure 2.17 below gives a pressure–temperature diagram of the fixed mass of water contained in the cylinder. The straight line AB plots the cooling of the water mass as we have just described it, with the pressure held constant. At the temperature and pressure associated with point A, only vapor is present. As the temperature decreases with cooling, the line moves first through the liquid phase and then reaches the solid phase at point B.

The curved line separating the vapor and liquid phases defines the dew point, the points at which liquid water—dew or condensation—forms out of

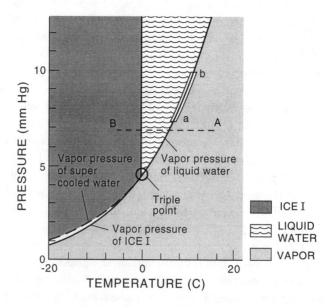

FIGURE 2.17. Phase diagram of water in the region of the triple point. The triple point is at a pressure of 4.5 mm Hg and a temperature of 0.0075 °C. The excursion reading from right to left along the line A to B occurs under cooling at a constant pressure. Excursions along the curved line ab in any direction along a phase boundary in each phase determine the pressure–temperature relation for that boundary.

water vapor onto a surface at the different combinations of temperature and pressure plotted on the curve. For the constant-pressure process described above, the dew point is the temperature where the straight line AB crosses the phase boundary from vapor to liquid; with varying pressure, we observe varying dew points. Also interesting is the fact that the vaporization point of water rises with increased pressure, and the freezing point decreases slightly with pressure. Water is almost unique in this instance; for most substances, the melting or freezing point rises with increased pressure.

Now, notice that at the point where the vapor–liquid and liquid–solid curves meet, they coalesce into a single curve. Below that pressure, the solid sublimes directly to vapor or the vapor sublimes directly to frost without going through the liquid phase. This junction point where the three phases can coexist is the *triple point*, where the pressure is 4.5 millimeters of mercury (mm Hg) and the temperature is 0.0075 °C.

A more precise discussion of these phase changes requires detailed reference to the principles of thermodynamics to derive the quantitative relations. Readers interested in a technical discussion of these processes should see Technote 1, page 192, "Thermodynamics of Phase Changes." Here we will continue our discussion by using a principle attributed to LeChatelier, which states that any system will try to relieve itself of the stress of pressure. Thus a given volume of gas expands to relieve any pressure upon it, if possible. Once again, let us consider the system given in Fig. 2.16 in which we have a cylinder fitted with a piston and filled with water existing in both liquid and vapor states. This time let us assume that the cylinder is thermally insulated.

The system is at some point a on the liquid–vapor curve given in Fig. 2.17. Compress the system by moving the piston into the cylinder a small amount, thereby raising the pressure by some amount ΔP; to relieve the pressure, water vapor will condense to liquid water, and the heat released in the process will raise the temperature of the system by some amount ΔT, moving to the point b. Thus the slope of this curve, $\Delta P/\Delta T$, is positive.

Suppose the cylinder had only ice and liquid present in equilibrium, as represented by some point on the ice–liquid curve. Upon raising the pressure in the cylinder, because the liquid has a higher density than the solid, some ice would melt, relieving the pressure. The heat required for melting, as given by the heat of fusion, causes the system to cool some amount ΔT, while the pressure increases by another amount, ΔP. Because the change in volume is much smaller in the case of the phase change from liquid to solid (freezing) than it is for the phase change from liquid to vapor (vaporizing), the curve is steeper, but, more importantly, the curve has a negative slope. That is, if ice is compressed, it will melt.

The liquid–vapor curve extends below the triple point in Fig. 2.17 as a dashed curve, which has a lesser slope than the ice–vapor curve because the heat of vaporization of liquid is always less than the heat of vaporization for the solid at the same temperature. So the vapor pressure over supercooled liquid is always larger than the vapor pressure over ice at the same temperature. Thus if droplets of liquid and particles of ice coexist in the atmosphere, the ice—or snowflakes—will always grow at the expense of the water droplets. The values given in Fig. 2.17 only apply for plane surfaces between phases; they must be modified when they are applied to cloud and snow formation in the atmosphere and to the changes we have discussed above that occur in the snow cover on the ground.

In our experience with skiing, we never come precisely to the triple point, which occurs at 4.5 mm Hg of pressure. At one atmosphere of pressure, or 760 mm Hg, the melting-point temperature is defined to be 0 °C—the melting point of ice; the triple point is at the temperature of +0.0075 °C. Below the triple point, the dashed curve in our diagram gives the vapor pressure over supercooled liquid water, which, as we have seen, is slightly larger than the vapor pressure over ice. Remember that liquid water will freeze to solid ice only when there is a means—a nucleation site—for the randomly oriented molecules in the liquid to line up like soldiers in a platoon. Otherwise chaos exists and the liquid never freezes. The shapes of the phase curves that join at the triple point are related to the quantity of heat that must be supplied to melt a unit mass of ice or to vaporize a unit mass of either ice or liquid water. The heat required to melt ice, commonly called the heat of fusion, is 80 calories per gram (cal/g); for liquid water at 0 °C, the heat of vaporization is 596 cal/g; and for ice at 0 °C the heat of vaporization is 676 cal/g. These heats will be important later when we examine the sliding of skis on snow.

Phase Changes and Curved Surfaces

In the discussion that follows, we consider a plausible, qualitative argument for describing how the phase changes occur for the curved surfaces of water droplets and ice crystals. Recall the shape of the hydrogen and oxygen atoms as they chemically bond to form a molecule of water (see Fig. 2.12) and the alignment of water molecules in ice crystals as they were represented in Fig. 2.15. The oxygen atoms on the plane faces of the crystals lack one bond; on the corners or edges of the crystals, some of the oxygen atoms may lack two bonds. Some of these open-bond sites are indicated by the stub connectors shown in the model.

In Fig. 2.18 we see a schematic representation of the open bonds that would occur at one of the plane surfaces of the ice crystal lattice. The

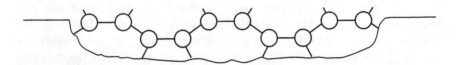

FIGURE 2.18. Schematic representation of an idealized arrangement of ice molecules in a plane, crystal surface such as was shown in Fig. 2.14. Only oxygen atoms are shown. Note the open bonds represented by the short stubs on the atoms along the top row.

density of available bonds is nearly uniform. At the molecular level the surface is not exactly plane, so the figure represents something of an idealized version of reality; however, the general effect is true enough.

Lacking the full complement of bonds, those oxygen atoms associated with water molecules on the corners or edges of the ice crystals—rather than on the plane surfaces—will be more easily dislodged from the crystal lattice by thermal agitation or external bombardment. Figure 2.19 illustrates the conformation of an ice crystal that one would find in a snowpack. The molecules form a convex surface if there is no material to the upper right, and a concave surface if there is no material to the lower left. Oxygen atoms

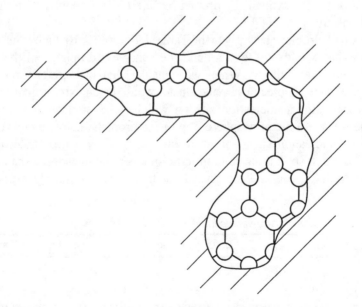

FIGURE 2.19. Internal lattice structure of an ice grain as it might appear in a snowpack.

on the convex surface will dislodge under molecular collision more easily than will oxygen atoms arrayed on a plane surface. Also, molecules will tend to stick and bond better on the concave surface than they stick and bond on the convex surface. Thus if all of the surfaces are at the same temperature, the vapor pressure will be higher at the convex surface and lower at the concave surface relative to the plane surface.

Another way of looking at this process is to consider the analogy of a water-filled balloon. The water in the balloon is under pressure, due to the surface tension created by the stretched balloon, just as the water in a spherical droplet is under pressure due to the stretching of its surface tension. That is why water tends to vaporize more readily from the droplet form than it does from a plane surface, by the principle of LeChatelier, which states that any system when under pressure will try to relieve itself of the stress of that pressure—in this case by changing state from liquid water to vapor.

Snow crystals and fog droplets coexist inside a cloud; they also coexist in snow cover on the ground where liquid water, ice, and water vapor may all be present in varying amounts. Relative humidity is the ratio of the water vapor pressure in the air to the pressure over liquid water at the same temperature. Table 2.1 gives the relative humidity, which is also called the saturation ratio, of the water vapor around a spherical water droplet of a given radius to the relative humidity that exists over a plane surface of water at 0 °C. The typical fog droplet has a radius of about 10 μm; so the saturation ratio is about 1.0.

The small droplet sizes given in Table 2.1 have 2–10 times the vapor pressure that would be found around the usual fog droplet, and thus they would evaporate quickly. When we consider that the aggregation of as many as 23 water molecules in one cluster would be an extraordinary event, the necessity for a nucleation site on which water molecules may stick becomes clear. In the atmosphere, fog or ice droplets must contain foreign particles—nucleation sites—of sufficient size to allow water molecules to bond with them so that the saturation ratio never becomes more than about 1.25—otherwise, fog and ice crystals in the air would simply evaporate.

TABLE 2.1. *Saturation ratio versus interface radius for liquid water.*[a]

Radius r_{lg} (μm)[b]	Saturation ratio	Number of molecules
∞	1	∞
1.7×10^{-3}	2	730
5.2×10^{-4}	10	23

[a]Adapted from data presented in Rogers [4].
[b]1 μm=4×10^{-5} in. lg=liquid gas.

This result suggests that no clouds—and no snow—could ever form in an absolutely clean atmosphere.

REFERENCES

1. For more on this subject, see J. Chen and V. Kevorkian, "Heat and Mass Transfer in Making Artificial Snow," Ind. Eng. Chem. Process Des. Develop. **10**, 75 (1971).
2. For more on the subject of making artificial snow using bacterial nucleates, see R. Fall and P. K. Wolber, "Biochemistry of Bacterial Ice Nuclei," in *Biological Ice Nucleation and Its Applications*, edited by R. E. Lee, Jr., G. J. Warren, and L. V. Gusta (APS Press, St. Paul, MN, 1995), pp. 63–83; R. J. LaDuca, A. F. Rice, and P. J. Ward, "Applications of Biological Ice Nucleation in Spray-Ice Technology," *ibid.* pp. 337–350. Our discussion of this topic is generally indebted to these sources.
3. An international classification system describes the varieties of snow found on the ground. It is used primarily by the various professionals who work with snow and avalanche control for ski areas, transportation departments, and so on. For more information on this subject, see S. Colbeck, E. Akitaya, R. Armstrong, H. Gubler, J. Lafeuille, K. Lied, D. McClung, and E. Morris, *International Classification for Seasonal Snow on the Ground* (International Commission for Snow and Ice, World Data Center for Glaciology, University of Colorado, Boulder, CO, 1990).
4. R. R. Rogers, *A Short Course in Cloud Physics* (Pergamon Press, 1976).

CHAPTER 3

EQUIPMENT
Properties and Performance

Although the modern ski is the result of perhaps 4000 years of evolutionary development, only in the last 50 years or so have skis, boots, bindings, and other equipment been consciously designed to operate together as a unit, and thus allow the lower leg to transmit to the ski the forces and torques necessary for precise control on steep downhill pitches. There are now a dozen or more different types of skis and boots designed for different types of skiing. By far the greatest number of skis and boots are produced for alpine skiing, which includes such varied types of skiing as downhill, giant slalom, slalom, and general recreation skiing. Each of these pursuits calls for skis with special features and characteristics. After alpine skiing, the general group of nordic skis designed for use on prepared tracks includes touring skis, diagonal stride skis, and freestyle skate skis. Cross-country skiers who venture off the prepared tracks may use free-heel skis, such as the telemark racing types, which are similar to some of the alpine racing models; general backcountry touring skis; or mountaineering skis designed to be used in untracked snow of any kind. Finally, consider that with very few exceptions the basic maneuvers used by snowboarders are much the same as those used by skiers, so a section of this chapter discusses the physical features of snowboard equipment.

The extensive selection of different ski makes and models serves to satisfy the equally various tastes, styles, and needs of individual skiers. In this chapter, skiers will discover more about the physical properties that affect the performance of the skis, boots, bindings, and other equipment they use (or are considering using). Readers may ask more informed questions of ski vendors; although, when they do ask such questions, they may find that many ski manufacturers do not provide vendors with much technical information. At best, when you know what to look for, you may better decide for yourself what equipment you need to make your skiing performance most satisfying. No matter how a ski may be specially designed for a specific use, all skis have certain general characteristics and specifications that we should look at first [1].

SKI GEOMETRY

Technical brochures available from most ski manufacturers list the dimensions of the skis they sell, but rarely are there sufficient details related to the skis' performance. In the discussions that follow, we consider the most important elements of ski geometry and discuss briefly the materials used to make skis and the actual process of their fabrication. Most of our attention, however, will focus on the dynamic properties of skis as they influence the skis' performance. For the discussions that follow, refer to Fig. 3.1 which illustrates the general geometry of essentially all ski types.

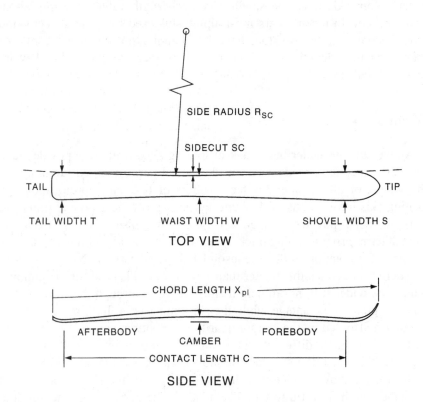

FIGURE 3.1. Geometric ski dimensions described using common nomenclature.

Length

The length of the ski, from tip to tail, is called the *chord length*, which we will designate as X_{pl}, where *pl* is the projected length. Metric measure is used worldwide for skis. Even people who have skied only once will likely remember that, when renting their first pair of skis, the shorter ones were 120s to 160s, the longer ones were 210s or 215s. Those lengths are measured in centimeters. In the nineteenth century, skis were as long as 3 meters (m); today, modern skis are usually 2 meters (200 cm) or less in length. If wary skiers take the time to measure the skis they see in the shops, they will find some differences among the manufacturers in the actual length of the skis relative to their stated lengths.

More important than the chord length is the contact length, C, which is also sometimes called the running surface length. The contact length is the portion of the ski that contacts the surface of the slope during normal performance. Contact length is important in determining the carving radius of a ski; a shorter contact length allows the skier to make a tighter turn. Thus slalom skis have the smallest contact-length values, running about 160–180 cm. In recent years most alpine skis used for general recreation have become shorter; so, too, have backcountry powder skis. A rule of thumb regarding length is that longer skis have greater stability, but they are more difficult to turn.

Width

Alpine skis are wider than track or touring skis, while true powder skis, with the exception of skis used for the special purpose of ski jumping, are the widest of all. Alpine skis have widths of 6–9 cm; nordic track and skating skis have widths of 4–7 cm. For skiing off the groomed trails and into the deep powder—sometimes called "off-piste" skiing from the French term, *piste*, meaning track—the extent of the ski width is dictated by the need for flotation in the unconsolidated, powder snow. New models of powder skis have widths greater than 10 cm. The skis used for ski jumping tend to be wider still, not to help them slide on or through the snow, but to take the greatest advantage while the ski jumper is in flight of the aerodynamic lift afforded by the extraordinarily wide bottom area on the ski. Note that skis have three different width values, each of which are illustrated in Fig. 3.1: the waist, the shoulder or shovel width, and the tail width.

Seen from above, notice that the width profile of the ski is not uniform over the length from tip to tail. The depth of the curved edge measured at the waist of the ski is called the *sidecut*. As we will see when we examine the geometry of the carved turn, the sidecut—this curved, lateral edge—determines the radius of the turn that can be made on a given pair of skis. The sidecut value taken with the contact length gives the radius of that turn, the sidecut radius R_{SC}. The sidecut is largest for slalom and recreational skis, ranging from 0.6 to 1.5 cm. The exact profile of the sidecut is really not all that critical, except for high-performance skis; nevertheless, some manufacturers tout the performance advantages of their special sidecut profiles. Understanding how the ski edge performs should help skiers judge for themselves the merits of one or another sidecut profile.

Skis also have a waist, created by the sidecut, which, for recreational skiing, is usually located at about the position where the boot is set in the bindings. The position of the waist ranges from the center of the running surface to some distance behind the center, a characteristic of slalom racing skis. For slalom racing, the boot is positioned behind the waist so that the resulting higher heel pressure helps the slalom racer release the tips of the skis. Mounting the boot behind the waist also shortens the edge running length of the skis, which shortens the skis' turn radius and helps the slalom skier make the quick turns needed to race from gate to gate. Skis also have a shoulder (more often called the shovel) width, which is the widest part of the ski's forebody. The tail width is the widest part of the ski's afterbody, and it is normally somewhat narrower than the shovel width.

Camber

A ski's camber is the curvature of the ski's running surface when it is under no load. Camber increases the ski's stability by controlling the boot load, which is transferred through the ski along its length to the tip and tail. A ski designed for off-piste skiing generally is soft in flexure and bends easily to create the reversed camber necessary for turning in some back-country snow conditions. Skis with high torsional rigidity—that is, skis that do not twist or torque much when one edge is set into the snow during a turn—do not need as much camber to be controlled well. More details on these subjects will be given later when we examine the physics of carving a turn.

Thickness

Ideally, skis intended for most uses should be as thin as possible while providing the structure needed to achieve the required bending stiffness and torsional rigidity. Exceptions to this rule are track or touring skis, which have a double camber, and alpine racing skis, where the binding–boot assembly must be raised far enough off of the body of the ski so as not to drag through the snow when carving on edge during a turn. In the latter case, a special platform may be installed on the ski under the boot and binding just to provide the necessary snow clearance. The platform, binding, and boot system interacts with the ski structure, which affects the underfoot bending stiffness of the ski and the behavior of the ski.

These are the essential geometric dimensions common to all skis that significantly affect a ski's performance. Most of the important performance characteristics of a ski are determined by the dynamic properties associated with these geometrical characteristics.

SKI MATERIALS AND CONSTRUCTION

Just as the original, single-piece, wood-slab skis gave way to laminated wood skis with metal edges, so the modern ski has evolved to be made of metal, fiberglass, and other composite materials. The availability of new materials with remarkable strength and tensile properties has made it possible to design skis for specific uses. Using curing resins for bonding and molding permits the development of skis with specific bending and torsional flexure patterns as well as specific vibration-damping features. The development of true, torsion-box or tubular construction for skis, as contrasted with a top-and-bottom, laminated arrangement of materials, has permitted the design of skis with much improved torsional rigidity, allowing the skier greater edge control. The core of a ski serves chiefly as a separator for the ski's structural elements. Cores may be injected in the molding process or fabricated from a light material. Laminated wood cores are frequently used, especially with layers of viscoelastic materials that enhance vibration damping. The steady evolution of construction capability has led to a corresponding design progression that has produced ever shorter and lighter skis that offer skiers softer bending stiffness combined with increased torsional stiffness, a pairing of qualities that promises marked improvement in the skis' performance over a variety of terrain.

The chief structural elements of the ski—the top and bottom plates or the tubular shell—may be made of steel or aluminum alloy sheets; of ceramic fibers, such as fiberglass or the composite fiber Nextel®, which is made of aluminum, boron, and silicon; or of organic fibers, such as carbon, aramid (Kevlar®), or polyethylene (Spectra®).

The physical properties of interest are density, tensile modulus (also known as Young's modulus), tensile strength, and the strain-to-failure limit (see Table 3.1). The tensile modulus is the constant of proportionality between stress, given by F/A (the force applied axially to the sample

TABLE 3.1. *Properties of high-modulus and high-strength fibers.*

Fiber		Density (g/cm³)	Tensile modulus (msi[a])	Specific modulus (msi[a])	Tensile strength (msi[a])	Price per pound
Metallic						
Hardened steel		7.85	30	3.8	0.29	
Aluminum alloy	7075-T6	2.75	10.4	3.8	0.075	$3.50
Aluminum–magnesium –titanium	Titanal	2.75	10.4	3.8	0.084	$4.50
Organic						
Carbon	Graphite	1.66	33–55	20–33	0.35–0.45	$18
Aramid	Kevlar	1.44	10–25	7–17	0.4	$11
Polyethylene	Spectra	0.97	17–25	18–26	0.38–0.43	$20
Ceramic						
Fiberglass	S. Glass	2.5	13	5	0.5–0.66	$3
Aluminum–borate–silicon	Nextel	2.85	22–33	8–12	0.25–0.3	$65

[a]msi denotes millions of pound per square inch.
[Reprinted with permission from *The Ski Handbook* (K2 Corp., 1991).]

divided by the cross-sectional area) and the resulting strain on the material measured by the elongation per unit of length, $\Delta d/d$. Thus the tensile modulus y may be expressed as follows:

$$\frac{F}{A} = Y \frac{\Delta d}{d}. \tag{3.1}$$

The limiting elongation in percent represents the material's strain to failure or deformation. For most of these materials, the modulus for tension and compression is the same, but their tensile and compression strengths may be very different. For example, carbon and fiberglass are usually much more fragile under tension than they are under compression.

For many years, sheet-metal materials were commonly used to make skis because of the ease of forming the material, even though the specific strengths of the metals are much lower than the strengths of organic or ceramic fibers. Today, because of advances in construction techniques and because weight is often a key design consideration, organic and ceramic fibers with high specific strengths have largely replaced metal, even in less expensive skis. The majority of today's ski designs use resin-impregnated, unidirectional fibers that are laminated around the ski core in strips, diagonal weaves, or triaxial braids. Because fiberglass combines a large strain-to-failure value with low cost, it is the most frequently used structural element in making skis. Some fiber–composite skis may suffer from internal fiber failure under too much repetitive flexural stress. Skis that exhibit this problem do not break, but their bending stiffness decreases markedly; the skis become "soft," and their performance deteriorates.

A variety of materials are used to make the top and bottom (or base) of the ski. Some of the more common materials are listed in Table 3.2. The materials used to make the top of the ski protect and beautify the ski. These materials must bond well to resins; must be impact, abrasion, and weather resistant; and must be capable of being decorated—they must hold paint or ink. The complex decorations found on modern skis require many silk-screening passes, so the inks used in this process must adhere well to the surface of the ski.

Ski base materials are much more functionally important than ski top materials, and they must meet some very special requirements. Studies of sliders on snow—which includes skis—have shown that the ski base must be abrasion resistant, as hydrophobic (nonwettable) as possible, yet it must also be porous enough to hold running wax well. At very low temperatures, the ice grains that make up the snowpack are like grains of sand, and they will abrade the surface of a ski base, as will dust or dirt on the snow. At all temperatures, frictional heat generates a film of water on the ski base that, while it lubricates the ski, should not uniformly wet (or soak into) the surface of the ski base.

The two materials with the smallest coefficient of sliding friction are polyethylene and Teflon. Teflon, however, is not tough enough to hold up well against abrasion. It is also expensive. Polyethylene, if it has a high density and is cross-linked sufficiently to make it tough but not brittle, works quite well. To make a ski base, polyethylene is granulated and sintered (pressed and heated to bond the granules) one or more times to make a sheet of material that will retain wax and be hard enough to be surface-ground to provide a microgrooved running surface. Graphite or carbon black is added to the polyethylene to make the material conductive, which decreases the triboelectric static charging effect on the bottom

TABLE 3.2. *Thermoplastic properties of materials used to form ski tops and bases.*

Thermoplastic material	Main constituents	Specific gravity	Coeff. of lin. exp. (in./°F)	Heat def. temp. (°F @ 66 psi)
ABS	Acrynotrite–butadiene–Styrene	1.05	$(5-6)\times10^{-5}$	210–225
UHMW	Polyethylene	0.94	8×10^{-5}	203
Triax	Polyester-nylon			235
Isotop PK	ABS urethane	1.03	6×10^{-5}	190
Nylon 11	Nylon	1.04	5.5×10^{-5}	302
Macroblend	Polyester, polycarbonate	1.22	4×10^{-5}	239
Vandar (Teflon)	Polytetrafluoroethylene		8×10^{-5}	

[Reprinted with permission from *The Ski Handbook* (K2 Corp., 1991).]

surface. In some cases, additives to the base material in the shovel region of the ski increase the formation of a water film that enhances the lubrication of the entire ski base. For this same purpose, the base material's optical absorptivity for visible radiation—that is, its ability to soak up heat from sunlight—may also be important for the heating it provides to the ski base material.

In modern designs, the ski core is basically a spacer—a form for positioning the top and bottom structural members of the ski—that gives the ski some degree of its necessary stiffness and strength and contributes to the ski's weight. The core may be subject to internal shear stresses, but these stresses are quite low. The most important role that the core material plays is damping vibration, which increases the stability of the ski. For that reason, wood cores are most frequently used in racing skis. Laminated wood cores are formed to the required specifications from a range of hard to soft woods: ash, birch, maple, hickory, fir, spruce, or poplar. Wood's ability to increase vibration damping and structural stability is a more important consideration than its relatively greater weight. The lightest cores are made of foams or honeycombs of synthetic materials. In these cases, the structure of the ski must be a true box in which the sidewalls support the shear and compressive stresses placed on the ski. In some fabrication procedures, the ski core is injected as a foam that expands, pressing the structural members against a forming die and forcing resin into the interstices of the composite materials used to make the structural members, thus enhancing the bond strength.

Finally, consider how high-performance skis require a careful blend of liveliness and vibration damping. To achieve this balance between stability and feel, manufacturers of high-performance skis use viscoelastic damping layers—thin elastomer layers inside the ski that offer damping and prevent excitations from resonating throughout the ski structure. Table 3.3 presents the characteristics of some common materials used to damp vibration in skis. Neoprene and urethane sheets are commonly used to provide damping. Many of the bonding resins used in ski manufacturing have viscoelastic properties that contribute to vibration damping. Skis may also use cracked or segmented edges to prevent the shock excitations generated in the edges from propagating into the ski structure and producing the effect known as *chattering*. In general, as skis of all types become lighter, softer, and shorter, the need for effective vibration damping increases. Although this is especially true for racing models, recreational skis increasingly also need some vibration damping because good performance depends on the ski maintaining its edge contact on packed snow.

TABLE 3.3. *Properties of viscoelastic materials used for vibration damping.*

Name	Density (g/cm^3)	Flex mod (psi)	Uses
Thermoplastics			
ABS	1.04	310 000	Tops and sidewalls
Nylon 12	1.02	180 000	Tops and alloyed with ABS
UHMW polyethylene	0.93	120 000	Base material
Elastomers			
Urethane	1.18	800	Inlays, top and heel protection
Surflyn	0.96	36 000	Sidewalls and inlays
Polyamide	1.01	560	Inlays, heel protector
Thermosets			
Epoxy resin	1.16	400 000	Fiberglass and carbon matrix

[Reprinted with permission from *The Ski Handbook* (K2 Corp., 1991).]

MECHANICAL AND DYNAMICAL PROPERTIES

The important physical parameters for understanding how a particular ski behaves are vibration-damping properties and the distribution of the flexural and torsional rigidities. The properties of the base materials—the running surface preparation, its composition, and the thermal properties of the ski body—are important for racing and other high-performance skiing. How a ski tracks, carves, or chatters and how much attention the skier must give to keeping the ski under control depend to a great extent on the ski's vibration-damping properties and flexural and torsional rigidities.

For modern ski techniques, the properties that most affect ski performance are the side geometry of the ski and the distribution of flexural and torsional rigidity over the length of the ski. Longer skis have greater stability and track better at high speeds when their length is coupled with forebody and afterbody bending stiffness. When these properties are combined with a small sidecut or a large sidecut radius and small taper, the ski becomes a downhill or a giant slalom, high-speed ski, respectively. The sidecut radius is the radius of the circle that can be defined by the shovel, waist point, and the tail of the ski. The sidecut radius is given below by the following relation, which is an approximation for the radius of a circle passing through the tip, tail, and waist points of the ski edge:

$$R_{SC} = \frac{C^2}{8\ SC},\qquad(3.2)$$

where C is the contact length; the sidecut, SC, is given by the relation SC

$= \frac{1}{4}(S - 2W + T)$, where S is the shovel width, W is the waist width, and T is the tail width.

Taper is another factor that affects the turning stability of a ski. This quantity is defined as $\frac{1}{2}(S - T)$. A ski with a wide shovel tends to track with the forebody, while a ski with a wide tail tends to track with the afterbody. Downhill skis generally have large tapers; slalom skis generally have small tapers, but large sidecuts. A narrow-waisted ski will be easy to set on edge to initiate a carved turn and at the same time transfer load to the tip and tail. The interaction of a ski's edging ability with its bending and torsional flexure determines the distribution of the boot load along the ski. Skis with a large camber tend to transfer boot load to the tip and tail of the ski; thus a soft-flexure ski that is easy to bend may not carve as well as a stiffer ski. As we noted above, however, the combination of a soft flexure with a stiff torsional rigidity will produce a ski that carves well without much skid.

The overall mass of the ski is, in itself, not very important. The mass of the ski is usually the consequence of other design choices. Large mass is advantageous for skiing at high speeds because the ski's momentum, which is equal to mass times velocity, goes up with mass. Thus a given random force causes smaller changes in the velocity as the mass increases, so the ski holds its speed and remains stable. Although the added stability of heavier skis would seem to be a favorable quality, in general, the mass of most recreational and sport skis manufactured today has decreased as new materials and construction techniques allow the production of lighter skis that are easier for less-experienced skiers to control.

Swing weight is the mass parameter that most directly distinguishes the comparative performance characteristics of heavy skis and light skis. Swing weight is the moment of inertia about a vertical axis drawn through the boot. As mass is to linear motion, so the moment of inertia, or swing weight, is to rotational motion. A high swing weight is achieved by placing additional mass at the tip and tail of the ski. Such a ski has high rotational inertia, so random forces cause a smaller rotation or change in the direction of the ski, making the ski more stable. Conversely, a low-mass ski, while it is less stable, may be more easily manipulated by the skier.

Ski Flexure, Torsion, and Setting an Edge

A ski's bending stiffness determines how that ski transfers boot forces to the snow; that is, when we set the ski's edge into the snow, does the ski carve, skid, or chatter? Seen as a physical entity, the ski is a nonuniform beam loaded by the boot; the reaction force of the snow is distributed along the ski's bottom surface, as we see in Fig. 3.2. The ski boot transfers to the

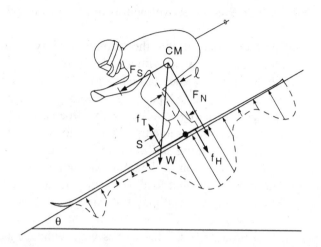

FIGURE 3.2. Forces acting on the ski through the boot.

ski a moment of force applied by the skier's leg and foot. Leaning forward or backward effectively moves the point of application of the normal force applied by the boot. The normal component of the skier's weight, F_N, is transmitted by the leg to the boot to load the ski. A pair of equal forces, f_T acting at the toe and f_H acting at the heel, generate a moment, $M = Sf_T$, where S is the heel-to-toe separation. This moment also equals ℓF_N. a skier may change the effective point at which the force F_N is applied by leaning forward or backward. The reaction force of the snow is always the sum of the distributed forces from the snow, but the distribution of the snow reaction force changes with the point of application of the normal force applied by the boot, F_N. The force F_S (the component of gravitational weight W parallel to the slope) is the amount of force that acts against inertia and drag forces to keep the skier in a state of motion.

Lateral moments are significant because, when we initiate a turn, forces are applied at the edge of the ski whether we turn by carving or by wedging, and the boot must transmit the required lateral moment to the ski from the lower leg. In Fig. 3.3, the load force F and the reaction force R produce a lateral, twisting moment M on the boot, which is transferred to the ski.

The ankle cannot transmit lateral moments as easily as it transmits fore and aft moments, so the boot must provide the necessary stiffness that allows the lower leg to transmit the torques to the skis that are needed to set the edges into the snow. The role that boot stiffness plays in controlled edging accounts for much of the difficulty that free-heel skiers have when they use conventional cross-country shoes or boots, which have little or no lateral stiffness, to ski on anything other than prepared tracks.

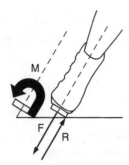

FIGURE 3.3. The lateral moment generated when the ski is set on edge.

To find the load distribution along the bottom or the edge of a ski when it is forced into a given snow surface configuration, one must consider the problem of a nonuniform beam in both flexure and torsion. The ski, like a bridge or a joist, is built with top and bottom structural members separated by spacers. The stiffness of the ski—its resistance to bending—depends on the separation of the member stressed in tension (the bottom) and the member stressed in compression (the top). The bending stiffness of a section of a ski is measured by the radius of curvature when a bending moment is applied at the opposite ends of that section. Readers interested in a more detailed exposition of this problem should see Technote 2, page 199, "Ski Loading and Flexure on a Groomed Surface."

The distributed property of the ski that determines local bending for any load distribution is the quantity called the bending stiffness, B. If any section of a ski is forced to bend by the application of moments at the ends of that section, the radius of curvature, R, is related to the applied moment M by the relation

$$M = \frac{B}{R}.$$ (3.3)

The bending stiffness of a ski, which is a composite beam, is a function of the width and thickness of the beam and the elastic moduli of the materials used in its manufacture. A ski's stiffness is determined, for the most part, by the use and placement of manufacturing components that have large tensile moduli, such as steel, aluminum alloy, fiberglass, or carbon–boron filament materials.

The overall stiffness of a ski is usually measured by supporting the ski at the shovel and tail contact points (which determine C, the contact length) and then placing a load at the midpoint, $C/2$. The stiffness, or center spring constant, is given by F/d, where F is the load and d is the center displacement. The stiffness for a beam that is a uniform, homogeneous bar, is given by

$$\frac{F}{d} = \frac{48B}{C^3}.$$ (3.4)

F is an arbitrary load placed at the center; d is the resulting displacement at that point. For a uniform beam, the midpoint deflection depends on the uniform stiffness of the beam and the distance C between the support points. Skis, although they are not uniform beams, usually behave similarly. For the typical ski, we would expect to find an effective bending stiffness B value of about 0.65–0.8 that of a uniform beam that has the same B modulus as that found at the center of the ski.

The center spring constant, or stiffness, of the ski determines much of its performance character, so the stiffness of the ski must be matched to the weight of the skier once the length of the ski is selected. In general, heavier skiers will prefer stiffer skis; so too will high-performance skiers. As we will discuss in greater detail below, the load distribution along the ski's edge is fixed by the ski's stiffness distribution. That part of the load that is not directly under the boot must be distributed evenly along the edge of a ski for the skier to be in control. A stiffer ski promises greater edge control.

Refer to Table 3.4, which lists geometric and dynamic properties for several different types of skis. The Head Comp and the Dynastar Omeglass are older models of racing or high-performance skis. The K2CVC and the Volant FX1 are contemporary high-performance skis, the modern successors of the first two skis listed. The K2-4 and the Elan SCX Parabolic are wide-body, radical sidecut skis that have been designed specifically to promote maximum ease of carving turns on groomed slopes. The Kneissel Ergo AVT and the S Ski Performer are also radical sidecut skis whose relatively soft flexure enhances their use in soft snow or powder. The Tua Excalibur is a relatively modern telemark ski.

Ski manufacturers do not normally provide such an extensive range of information about their skis' properties, even though that information can help skiers select skis suited to their particular style of skiing. For example, skis with a light swing weight may more easily be rotated to initiate a turn, making the swing weight one factor to consider when choosing a ski that will be easier to turn. The swing weights listed above are for a single ski rotated about an axis vertical to the plane of the ski and through the center of mass. Similarly, the balance of the individual spring constants for the forebody and afterbody is relevant for choosing slalom and telemark

TABLE 3.4. *Geometric and dynamic properties of some selected skis.*

Ski make and model	Head Comp	Dynastar Omeglass	K2 CVC	Volant FX-1	K2-4 88	Elan SCX Parab.	Kneissl Ergo AVT	S Ski Performer	Tua Excalibur
Chord length X_{p1} (cm)	207.0	187.0	198.0	187.0	188.0	201.0	179.0	184.0	186.0
Contact length C (cm)	180.0	169.0	178.0	165.0	168.0	181.0	156.0	164.0	165.5
Shovel width S (cm)	8.79	8.53	8.38	8.51	9.76	11.52	10.00	11.02	8.99
Waist width W (cm)	7.12	6.85	6.43	6.48	6.45	6.00	6.22	6.29	6.90
Tail width T (cm)	7.87	7.62	7.80	7.49	8.66	10.34	10.00	8.99	7.88
Sidecut SC (cm)	0.60	0.61	0.83	0.76	1.38	2.47	1.89	1.86	0.77
Sidecut radius R_{SC} (m)	66.9	58.3	47.7	44.8	25.6	16.6	16.1	18.1	46.3
Projected area A_p (cm^2)	1506	1306	1292	1176	1475	1532	1454	1578	1435
Pair mass M (kg)	4.99	3.45	3.74	3.97	3.53	4.09	2.72	3.02	3.17
Forebody spring C_f (N/cm)	20.4	18.2	17.2	19.2	21.2	20.0	15.2	14.4	17.9
Afterbody spring C_a (N/cm)	24.3	18.2	19.5	18.6	17.4	20.7	19.2	11.6	15.5
Center spring C_c (N/cm)	48.9	36.4	36.3	37.9	37.7	43.9	34.3	28.7	36.1
Forebody torsion T_f (N m/deg)	2.82	1.47	2.32	1.29	1.42	2.56	1.85	1.03	1.25
Afterbody torsion T_a (N m/deg)	3.47	1.34	3.47	1.58	1.57	2.41	2.35	1.42	0.97
Swing weight I (kg m^2)	0.65	0.48	0.52	0.49	0.48	0.59	0.52	0.41	0.40

skis, which need stiffer afterbodies to enhance their turning characteristics. These constants are measured by clamping the ski at the center and measuring the deflection when the tip or tail is loaded. Most skis have balanced tip and tail spring constants.

Let us consider (see Fig. 3.4) the bending stiffness in connection with the bottom load distribution for four of the skis listed above in Table 3.4. Note that, for clarity in reading the graph, the K2CVC curve is displaced upward by 0.05 units, and the curve for the Head model is displaced upward by 0.1 unit. Notice how much smaller the bending stiffness distribution is for the S Ski model, which indicates that it is a softer ski designed for powder skiing. Note that only the Head model has a maximum bending stiffness behind the midpoint of the ski, which accounts for the large difference in the forebody and afterbody constants listed in Table 3.4 for that ski.

The Head model is an older racing ski. It has a stiffer afterbody to allow racers to sit back on their skis and hold their edges without skidding coming out of a turn. High-performance skiers who want the feel of a stiffer afterbody and the enhanced turning characteristics that result may experiment

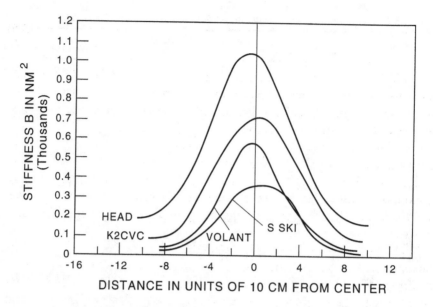

FIGURE 3.4. Bending stiffness distribution (*B*) as measured for four different skis. The curve for the Head model ski has been displaced upward 0.10 units. The curve for the K2CVC ski has been arbitrarily displaced upward 0.05 units.

with their equipment by setting up their skis to enhance the stiffness of the afterbody. Measure the forebody and afterbody stiffness separately and then move the clamping toward the tail to see how the stiffness of the ski would change with different boot placements. Ski with the boot placed at two different positions to get a feel for the difference that results in the skis' stiffness and turning characteristics. Because the stiffness of beams loaded by a point load uniformly varies as the length to the power 3, a 3-cm change in the boot mounting position for any of the skis listed in Table 3.4 would change the ratio of forebody to afterbody stiffness by about 25%. Skiers who try this experiment should feel a marked difference in how their skis turn and hold their edges on hard-packed snow.

A ski's lateral stiffness is 10–20 times greater than its bending stiffness and is of little concern. Torsional stiffness, however, must be matched with bending stiffness to give a ski its desired performance. Bias-ply fiber structures or rectangular tubular configurations achieve the highest torsional stiffness. Generally, the ski designer's goal is maximum torsional stiffness. Torsional stiffness is measured by clamping the ski at the center and measuring the ratio of torque τ applied to the tip or tail to the angle of twist, ϕ, in degrees. The expression relating the torque to the angle of twist is

$$\frac{\tau}{\phi} = \frac{2G}{C}, \tag{3.5}$$

where C is the length between the tip and tail contact points. The torsional stiffness modulus G is constant for a uniform bar. For a complex structure like a ski, G is the effective torsional stiffness coefficient, which will vary, as does B (the stiffness coefficient), along the length of the ski. The measurement described above gives an effective value for G. These values are commonly between 0.8 and 2.0 newton-meters per degree (N m/deg) (7–18 lb in/deg). When we discuss the distribution of force along the bottom of a ski, we need to refer to the specific values of the ski's flexural and torsional stiffness. Again, for more detail on these values, see Technote 2, page 199.

Some recent ski designs have introduced innovations that decouple the flexural and torsional stiffness modes of the ski. Advanced box or tubular construction techniques give the ski a much higher torsional stiffness coupled with the same degree of bending stiffness that one would expect from a similar ski using conventional construction. Another design innovation is the manufacture of skis with asymmetric cross sections. The inside edge section of such a ski has the larger depth or thickness, and the ski tapers to a much smaller value at its outside edge. Similarly, racing skis may have enhanced flexural rigidity at their outside edges. All of these modifications decrease the twisting moments along the ski and thus enhance edge control.

The base of the ski and the edges are essential for proper ski behavior. Metal edges are incorporated into all skis except nordic track models, and they are essential for modern ski technique. The steel edges bond to the bottom structural layer of the ski, contributing to the ski's bending stiffness. The steel edge may either be a continuous strip of metal or be partially segmented, so that it does not affect the ski's bending stiffness when the ski is in reverse camber, bending to contact a deformity (such as a mogul) on the slope. Because ski edges are dressed, that is, ground or filed, quite frequently, they must be several millimeters wide and thick. Usual base preparation calls for an edge base bevel of about 1° from the flat; the side bevel will usually be at 90°, but for better performance, the side bevel may be filed to yield a corner angle of about 86° to improve the edge bite on hard snow. The sharp edge is dulled with a file and sandpaper at the shovel and the tail from 5 to 10 cm inside the snow contact points for the inside edges and from 10 to 15 cm inside the snow contact points for the outside edges. This dulling allows the edge of the ski to release from the snow more easily to initiate a turn.

The base of the ski also contributes somewhat to its stability. All older model skis had a longitudinal groove in the base to give them tracking stability on flat surfaces. Skis used for jumping may have three or more such grooves in the base because these skis are not meant to be set on edge and turned: ski jumpers need stability as they gather momentum, skiing straight down the jump, riding, for the most part, flat on the base of their skis. Diagonal stride nordic track models as well as many backcountry models also have a longitudinal groove in the base to provide increased tracking stability when the skis run flat on the snow. Modern alpine skis, however, have no such groove. Instead, the polyethylene base of an alpine ski is ground with fine microgrooves that, after waxing, are wiped or buffed clean. This treatment is intended to give the flat ski the additional stability it needs. In racing applications, these microgrooves are longitudinal; recreational alpine skis are ground with the microgrooves in a crosshatched pattern to permit the ski to slide more readily sideways during turn initiation. Note that the base of the ski plays no role whenever the ski is set on edge in the snow.

Good and Bad Vibrations

As a rigid body, the ski has numerous natural modes of vibration. The flexural modes of vibration are the most significant, but the torsional modes also affect the edge chatter. Skis are subjected to impulsive forces that can excite all of the modes of vibration. For this reason, as we have seen, ski manufacturers incorporate materials into the construction of the ski that damp these vibrations internally; or, if necessary, external vibration dampers may be mounted on the ski. Lively skis transmit vibrations to the skier, who, through biomechanical feedback, responds to those vibrations—the "feel"—of the skis. An overdamped ski feels dead or sluggish; the skier receives no signals from the ski from which to fashion a response. Basically, different skis "feel" different to different skiers. A good vibration, or feel, for one skier may be a bad vibration, or feel, for another. An underdamped ski will vibrate too much and chatter, or, worse, release the edge entirely, so that the ski is out of the skier's control. The designer's task is to satisfy the preferences and ability of different skiers by crafting a fine balance from among the various vibrational modes that range between too much and too little damping. Generally, more competent skiers prefer more lively skis. Normal vibrational mode structure and damping is probably the most technical aspect of the ski design process. The biomechanical feedback loop comprised of the skier, the boot, and the ski subjects any analysis of this topic to the special needs and feel of the individual skier, so vibration damping is somewhat less a science, somewhat more an art.

A ski is clamped at the middle to the boot, forming, in effect, two almost independent, nonuniform beams—the forebody and the afterbody of the ski—that vibrate in the bending modes related to the ski's bending stiffness [2]. Torsional modes of vibration have natural frequencies outside the range for excitation of flexural modes. A mass on a spring has a resonant frequency ω given by

$$\omega = \sqrt{\frac{k}{m}}, \tag{3.6}$$

where k is the stiffness constant of the spring and m is the mass. The forebody and afterbody of a ski resonate in a similar manner, but with several different harmonic modes. For a ski driven at the center of mass, the distribution of the vibrational frequencies is shown in Fig. 3.5.

For most skis, the first mode has a frequency at 16–20 hertz (Hz or vibrations per second), the second mode is at 35–40 Hz, and the third mode is at 65–70 Hz. The torsional vibrational modes lie at frequencies above 100 Hz. Damping features present in a ski as a result of its bonding resins

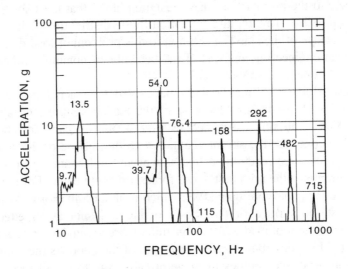

FIGURE 3.5. The frequency distribution of the normal vibration modes of a ski. The ski is clamped at the center to a shaker and driven. An output accelerometer located on the afterbody records the vibration response shown. [Reprinted with permission from R. L. Pizialli and C. D. Mote, Jr., "The Snow Ski as a Dynamic System," J. Dynamic Syst. Meas. Control, Trans. ASME **94**, 134 (1972).]

or wood core, or the use of viscoelastic sheets are tailored to give the ski the "feel" that the manufacturers sense is best for the particular group of skiers who are the market for that ski. Other properties being unchanged, damping shifts the resonance frequency to lower values, reduces the amplitude of the forced vibration that is excited, and then causes the vibration that does occur to die out. In practice, the amount of damping in a ski is fairly small because otherwise the ski becomes too dead. For special applications, tuned vibration dampers adjusted for relatively narrow frequency ranges can be mounted on the ski at the antinode of a particular vibrational mode. These dampers will then selectively damp at best one or two vibrational modes. For low-speed skiing, which includes probably 90% of recreational alpine skiing, vibration damping is really not much of a concern; it is only an issue for sustained high-speed, high-performance skiing.

THE LOAD-DISTRIBUTION PROBLEM

The configuration of a ski—short or long, narrow or wide, thick or thin in whatever combination—serves two basic purposes. The first is to support the skier's weight against the compaction of the snow. The second is to provide stability fore, aft, and laterally when skiing over uneven surfaces. Short skis do not have the stability of longer models, but they are easier to maneuver, so the problem becomes designing a ski that is as short and as stable as possible. This requires attention to the distribution of load on the running surface of the ski. The solution to this problem of load distribution involves an engineering analysis of the behavior of nonuniform beams on elastic and nonelastic support beds.

Because alpine ski areas usually have well compacted and groomed slopes, we will discuss first the case of the loaded ski pressing against a rigid surface. The ski is assumed to have a built-in camber. The force required to compress the ski into contact with the surface of the snow will develop point force reactions at the shovel and tail contact points. If the boot position is at the midpoint, these forces are about equal. For most skis, these forces are not more than 20 N, which is about 5 lb. For a ski pressed flat against the surface of the snow so that it is not at all edged, the remainder of the foot load will be transmitted directly under the foot and be represented by a distribution over the length of the boot. As the underlying foundation of snow becomes more deformable, the load still peaks under the boot, but it also spreads out along the ski, fore and aft. The snow may deform elastically, that is, return to its original state, when the ski has passed; or, as is usually the case, the snow will be permanently deformed in a depressed track. Certainly when skiing in new snow or skiing off-piste, the latter is always the case. First we will analyze the case of a ski pressed against a rigid, sloping surface.

For a ski resting horizontally on edge across the slope, the boot load will depress the ski until the ski makes contact with the snow surface over the length of the ski (see Fig. 3.6). The ski's sidecut determines the resultant curvature of the ski when the edge is bent to contact the slope [as it is in Fig. 3.6(a)]. Note that the three points, S, W, and T determine the radius, which is given by relation (3.2), of a cylinder that passes through those points and is normal to the plane of the ski. The intersection of that cylinder and the slope plane is the contact line of the ski depressed to contact the snow surface. From the geometry given in this figure, the effective sidecut of the arc shown in Fig. 3.6(a) is SC/cos Φ, so the radius for the circle made by the contact line, R_{con}, becomes

$$R_{con} = \frac{C^2 \cos \Phi}{8\ SC}.$$

(3.7)

This expression represents the radius of the circle that would be drawn on the slope by the contact line of the ski. Note also that the center of the ski is depressed by an amount SC tan Φ to the approximate arc of the circle formed by the vertical flexure of the ski, R_{flex}, for which the radius is

$$R_{flex} = \frac{C^2}{8\ SC\ \tan \Phi}.$$

(3.8)

If a ski is bent in flexure to some radius, R_{flex}, or it bends to fit the contour of a slope, a characteristic upward snow reaction force F_{reac} must appear

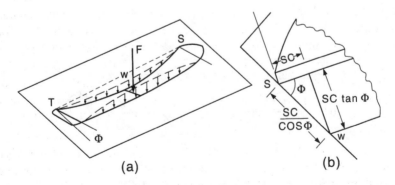

(a)   (b)

FIGURE 3.6. (a) shows a center-loaded ski deflected to contact an inclined plane. The dotted lines represent the unflexed ski. The arrows show deflection of the ski downward. (b) shows the geometry to determine the intersection radius of the edge with the inclined plane.

along the ski. For a ski bent to a constant radius of curvature over its length, the external loading per unit length along the edge is upward over the shovel and tail regions and downward under the boot. Figure 3.7 shows an example of such a distribution of the snow reaction force for the same Head Comp ski that appears in both Table 3.4 and Fig. 3.4.

To force the ski into a constant curvature over its full length, upward pressure is needed over the tip and tail, and downward pressure is needed over the midregion. Boot pressure is usually more than enough to bend the ski into contact with the snow; the remainder of the loading (which is not shown in the figure) is a uniform, underfoot, downward pressure that presses the ski into the snow.

The magnitude of the pressure needed for bending the ski is directly proportional to the curvature of the ski flexure. In equilibrium, the reaction load of the snow on the ski bottom is that shown in Fig. 3.7 plus a uniform

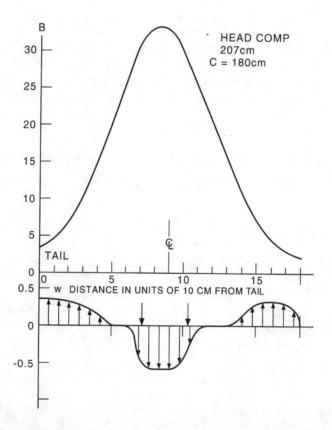

FIGURE 3.7. The edge loading distribution needed to deform the Head Comp ski to a constant radius of curvature.

upward pressure distributed under the foot, because the groomed snow surface is so hard that no additional flexure of the ski occurs. If a ski without camber were placed on a flat, rigid floor, the only pressure distributed on the ski bottom would be the force distributed directly under the boot. The load can be distributed along the length of the ski only through flexure of the ski. Note that the fraction of the total boot load that develops pressure at the tip and the tail of the ski is relatively small. The major part of the load cuts the snow surface under the boot to develop the support base for the skier. Unfortunately, the torsional twist of many skis at the tip and the tail caused by edge loading can make those skis perform 10–15% softer than they would if the skis' bending stiffness alone determined their softness. Additionally, a twist of the tip and tail of from 5° to 10° will cause a ski's edges to release, and the result is a skid turn rather than a carved turn.

When we discuss the mechanics of various ski maneuvers in subsequent chapters, we will focus more on the problem of loading distribution as it affects the ski's turning properties. For now, a good, qualitative feel for the load profile of a ski as it depends on the ski's elastic properties should be helpful, particularly when selecting a ski for purchase. Note, however, that most manufacturers, let alone retail outlets, are not prepared to provide much, if any, information on flexural or torsional rigidities as measured by the standard point-load tests we have used here. Unless you speak directly to a technical representative or a design engineer, you are not likely to get this information.

Figure 3.8 shows a sequence of qualitative load distributions on a bed of deformable snow for a variety of skis with a variety of properties. The bottom pressure profiles represent the total upward pressure from the bending of the skis as well as the pressure needed to support the skier load. The results presented in the illustration are qualitative only, meant to give the reader a feel for what happens in actual practice.

The graphed representation below each ski in the diagram represents the local reaction force of the snow bed on the ski, which deforms the ski into a reversed-camber configuration. The sum of these reaction forces must equal the total force exerted by the skier on the snow. For the sake of simplicity, we assume that in some cases the snow bed is groomed and incompressible and in other cases the snow surface is cut by the ski edge, as it would be in a skid. In general terms, the ski shown in Fig. 3.8(a) might be a recreational ski, while the ski shown in Fig. 3.8(b) might be a giant slalom or downhill racing model. The ski shown in case Fig. 3.8(c) might be a slalom or telemark racing model. The skis shown in Figs. 3.8(d) and 3.8(e) could be any of several types; the ski shown for Fig. 3.8(f) would be a soft, deep-powder ski that generates the large reversed camber needed for

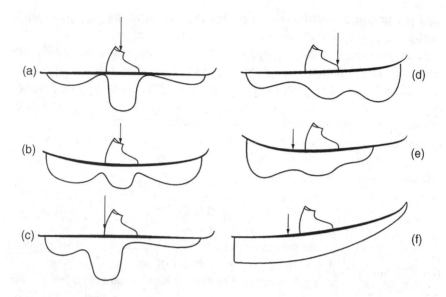

FIGURE 3.8. Bottom load distributions for six skis with different stiffness properties on snow beds of varied deformation.

turning in those snow conditions. All of the cases except (f) are for a packed and groomed alpine slope.

In Fig. 3.8(a) we assume that the ski is soft in flexure overall and symmetric in the degree of its forebody and afterbody flexure. This ski might be used for moguls or for general recreational skiing. The snow bed is extremely compacted; it suffers no deformation at all. A small, reversed-camber deformation from the edge-contact geometry yields some loading in the fore and afterbody regions of the ski, but most of the boot load is distributed directly underfoot to the snow in the region of the boot on the ski. The underboot loading would spread out along the ski if it were stiffer in the boot region or if the snow suffered deformation.

In Fig. 3.8(b) we see the case of a stiff, symmetric ski forced into a short-radius, circular carving turn. This stiff ski might be a giant slalom or downhill model. The loading in the fore and afterbody regions is increased markedly by the curvature of the ski, and the underboot loading decreases because the total bottom reaction equals the skier load.

The case represented in Fig. 3.8(c) is similar to the case illustrated in Fig. 3.8(a), but the ski pictured in (c) has a stiff afterbody. The relatively softer forebody and stiffer afterbody suggest that this ski might be a slalom or telemark model. Because a larger reaction force appears over the tail of the ski than appears over the front of the ski, the skier must rock backward to load the heel more than the toe of the boot, so the underboot loading on the ski moves backward as well.

In Fig. 3.8(d), the skier is in the stage of turn initiation in which the forebody ski edge is loaded to generate the groove for the ski edge to ride in through the turn. The forebody edge of the ski must be skidding across the snow surface, so the loading is markedly larger at the forebody than it is at the tail, where the ski slides directly in the groove in the snow carved by the ski's edge. More overall loading of the ski appears in this illustration because it represents a case at the initiation of a turn where the skier's speed is decreased and the large skidding force loads and deflects the forebody of the ski. For this case, the effective boot loading point must move forward of the toe. The ski represented in the illustration might have a soft afterbody and a stiff forebody. These characteristics would make the ski difficult to steer; consequently, skis would not be designed to have these characteristics. One might, however, find an older ski with similar characteristics if the afterbody of the ski had "broken down" and become "soft" from excessive use while the forebody retained most of its original stiffness.

In Fig. 3.8(e) we see the case in which the skier reaches the conclusion of an aggressive turn and uses lateral projection—the skier pushes off of the outside carving ski to transfer loading to the opposite ski and initiate the next turn. The afterbody of the projected ski becomes heavily loaded; sometimes the forebody may not contact the snow at all during this maneuver. Thus the afterbody of the ski is curved by loading while the forebody is almost not deflected at all. The boot loading point moves aft of the boot heel.

In Fig. 3.8(f) we see the case of a ski designed to be used in soft, uncompacted snow. This ski has a rather soft flexure and is able to generate a large reversed camber. It will initiate turns and carve well, but it is less stable at high speed. The reversed camber of the ski results from the distributed, upward compaction pressure on the ski as it slides through the snow. This is the case treated in detail in the discussion given in Technote 12, page 244, "Ski Flexure in Uncompacted Snow." The geometry illustrated here shows that the tail of the ski slides forward flat in the track; as soon as the ski slides other than flat in the track, it will plow the snow and increase the drag friction, bringing the skier to a halt. The bottom force distribution depends on the depth to which the ski sinks into the snow. The tip of the ski must always run close to the snow's surface, so the ski bends as shown, with its afterbody much more loaded than its forebody. Hence the effective loading point of the boot moves backwards. The skier does this by sitting back on the ski with knees bent, so that a moment is applied at the boot, the toe is lifted up, and the heel is depressed, lifting the tip of the ski and loading the afterbody of the ski.

BINDINGS AND BOOTS

Ski bindings have evolved from a simple toe fixture to a toe piece with a heel strap to still more complicated fixtures that allow the skier to transmit the forces necessary for control of the ski through the boot and to the ski. As soon as lift facilities became widely available so that level or uphill travel was no longer necessary, the equipment we today call "alpine" developed rapidly. Free-heel boots and bindings stayed more or less the same in their design for some time while a rapid evolution occurred in alpine boots and bindings. The evolutionary pattern of boot and binding equipment has been circular, moving from the original free-heel gear to alpine gear with its safety bindings and boot developments, and, now, in recent years with the increased popularity of telemark and backcountry skiing, back to free-heel bindings that incorporate some of the features of fixed-heel bindings, including safety-release systems.

As skis have evolved to become lighter, binding and boot systems have evolved to become heavier. The boots and bindings that most skiers use on the slopes today usually, when taken together, weigh appreciably more than the skis to which they are attached. Free-heel, backcountry gear is an exception to this trend because performance, which usually means the added feature of safety-release systems, is often sacrificed to achieve lesser weight. Free-heel, backcountry ski bindings with safety-release features are available, but they are not found much on general trail, track, or backcountry ski equipment. True mountain touring gear is the exception to this practice because the fast descent of steep pitches makes the addition of safety-release mechanisms necessary.

An early safety-release binding for alpine skiing, called the Cubco, was developed in the United States, some 50 years ago. The Cubco binding was one of the first that had full sideways and vertical release capability for both the toe and the heel. Today few modern bindings have a vertical toe release. The modern alpine binding incorporates an adjustable lateral and vertical heel-release mechanism with an adjustable sideways toe release.

With the advent of the extensive use of rental alpine skis, binding and boot systems must match and their adjustments must be calibrated to meet specified standards in order to avoid the potential liability associated with improperly matched or adjusted ski, boot, and binding systems. For this reason, some boots cannot be used with some rental skis and their bindings; however, the "footprint" of most ski boots is fairly standardized so that, for the most part, matching a boot to a binding is quite simple. For alpine bindings, the heel-to-toe spacing of the binding must be adjusted to match the boot size, and the height of the toe clips must be adjusted to accommodate variations in the sole thickness of the boot. Once this is done, the boot is bound to the ski and the safety-release mechanisms may be adjusted.

Alpine Boots: Fit and Feel

The boot is the steering wheel for the skis. As we have seen in the previous discussions and will analyze later in greater detail, the skier reacts to gravity and to acceleration and transmits force to the ski through the boot to create the geometric configuration desired for each ski, applying the forces and mechanical moments needed to maintain dynamic control. The largest forces are transmitted through the soles of the feet, so alpine ski boots must conform to the size and special anatomy of an individual's foot and leg to achieve the desired fit and feel.

As long as the ski boot extended to ankle height but no more, which was generally true of boots until the 1970s, no accommodation for the shape of the lower leg was required. Early boot designs stressed a snug, even tight, fit about the foot and ankle, because the leverage of the lower leg was not used that much in early skiing. Ski boots became notorious as uncomfortable monstrosities. The traditional leather boots gave way to hard, plastic shells that were stiffer and molded with a prescribed forward lean. The upper part of the boot was extended upward toward the calf, and the foot was fitted with insoles to give a proper fit that would be comfortable yet snug enough to transmit the appropriate forces and moments. Today, all high-performance boots and most recreational models reach well above the ankle to the calf or more, and they employ a variety of devices that allow them to be configured to suit the comfort and performance needs of the individual skier.

The design problem for alpine boots is to fit the individual comfortably with a boot that permits the articulation of the ankle and foot in a shell that is at once stiff enough to provide support and supple enough to allow the skier to feel and react to changing conditions. To accomplish this, alpine boots have a variety of ways to adjust and control the tension points. The basic flexure of the boot is the forward–backward angulation of the lower leg, which permits the upper body to remain upright while the effective center of loading on the ski is moved fore and aft by load distribution on the feet or by moment application. The hinge point for this longitudinal flexure should be positioned close to the ankle.

Free-Heel Boots for Telemark Turns

One of the authors of this book first learned to ski with free-heel equipment, and for years he carried duplicate sets of alpine boots and skis and free-heel touring gear until he recognized that a proper set of modified alpine skis, boots, bindings, and poles would serve for skiing both alpine and backcountry slopes. The essential point is that the heel of the ski boot must be free from the ski.

The early free-heel binding consisted of adjustable toe pieces with a leather heel strap, which soon gave away to a spring-loaded heel cable. Sometimes a rubber strap ran from the ankle to a clip on the ski positioned behind the boot. Then side clips for the cable were mounted at the edges of the ski to anchor the heel. Soon a tension release on the cable throw was introduced. The free-heel telemark binding that is widely used today is almost an exact copy of the early cable bindings used, for example, by U.S. Army mountain troops in World War II.

Let us look at the requirements of the boot–ski system for free-heel skiing. As in alpine skiing, the binding must rigidly fix the boot to the ski so that the boot–ski alignment is preserved and the leg forces and moments may be transferred to the ski with little or no play in the connection. The lightweight, three-pin binding with a soft boot that is commonly used for cross-country skiing on prepared tracks is not rigid enough to edge the ski satisfactorily. Some binding systems use a flexible heel plate to fix the boot to the ski. In most cases, a heavy-duty, anchored toe piece on the ski receives a boot that features the standard, 75-mm touring boot toe shape, creating a precise, angular boot–ski orientation.

The preferred scheme for anchoring the boot to the toe piece is a spring-loaded heel cable. The boot is flexible behind the ball of the foot, so the ball of the foot may be in contact with the ski to load and fix the edging of the ski, but the heel may be lifted free. A rigid-soled boot, pinned to rotate at the toe so the heel may lift free, is basically unstable. Either the knee comes forward until it hits the ski or it comes backward so the heel contacts the ski. These systems have no safety-release mechanism except for the cable spring system itself, which permits the semirigid sole of the boot to flex and lift out of the toe piece. For the skier who needs a surer release mechanism, a subplate mounting for the toe piece can be designed to release under torque in the manner of the early safety releases used on alpine bindings.

The free-heel, boot-binding system must have the same lateral stability and angulation control of the alpine boot. The binding must allow the skier to apply a backward moment to the ski, moving the load point toward the tail. Only rarely does the recreational skier have to apply a forward moment to the ski to move the load point forward. More often skiers shuffle a ski forward to gain forward–backward stability. Over the last few years, boots designed for telemark, free-heel skiing have featured a marked increase in their lateral stiffness. In some cases, plastic shells have been employed to enhance the stiffness of the boot to allow the skier increased edge control for carving turns on hard-packed snow. For off-piste skiing in soft or "crud" snow in which the ski is loaded uniformly to a reverse camber, increased lateral stiffness of the boot is not so crucial. Some backcountry skiers carry separable, plastic uppers that they apply to their boots for use on descents only. Finally, the telemark boot should, unless it is used exclusively at ski areas with lifts and groomed slopes, be constructed of materials that suit it for walking or climbing: the boot should be light and be fitted with lugged soles.

SNOWBOARD EQUIPMENT

The popularity of snowboarding, an offshoot of skiing that started about 1960, has expanded very rapidly in the decade of the 1990s. Today the snowboard enthusiast can choose from an array of snowboard makes and models that compares favorably to the variety of ski equipment available. Except for the way snowboarders propel their boards over flat ground—by removing a boot from the binding and pushing off in a manner identical to the propulsion of a skateboard—snowboard maneuvers when going downhill are quite similar to those of skiing [3].

There are roughly three classes of snowboards: the alpine racing or slalom board, the freestyle board, and the half-pipe, freestyle board. The freestyle snowboard is used for specialized snowboarding that is similar to freestyle aerial and mogul skiing, so we will not consider it in this discussion. Table 3.5 summarizes some of the geometric characteristics of snowboards.

Snowboards range in length from about 120 to 175 cm; they have contact lengths from about 90 to 140 cm and widths from about 20 to 30 cm. The snowboard's width allows snowboarders to plant their boots diagonally across the snowboard, which yields a projected area somewhat larger than that of a pair of skis. The important geometrical difference between snowboards and skis is the sidecut radius. Looking back at the dimensions for the skis listed in Table 3.4, note that the sidecut radii of the skis listed range from 67 m to about 45 m, with the exception of the narrow-waisted, radical sidecut models, which were specially designed to have very short sidecut

TABLE 3.5. *Geometric properties of snowboards.*

Board make and model	Burton Air 3.0	Burton Kelly Air	Burton Slopestyl	Burton PJ6.3
Overall length X_{pl} (cm)	127.5	161.0	171.0	161.0
Running length C (cm)	90.0	116.0	128.0	138.0(T)[a]
				139.0(H)[a]
Nose width N (cm)	23.7	28.8	29.2	27.1
Waist width W (cm)	20.5	25.0	25.0	22.3
Tail width T (cm)	23.7	28.8	29.2	27.9
Sidecut SC (cm)	1.6	1.9	2.1	2.5(T)[a]
				2.7(H)[a]
Sidecut radius R_{SC} (m)	6.3	8.8	9.8	9.5(T)[a]
				9.0(H)[a]
Projected area A_p (m^2)	2396	3833	3840	3718

[a]T=toeside, H=heelside for asymmetric snowboards only.

radii. All of the snowboards listed in Table 3.5 have sidecut radii under 10 m. Note also that all of the skis listed in Table 3.4 have the same sidecut radii on their inside and outside edges, but that one of the snowboards listed in Table 3.5 has a shorter heel side, or back, sidecut radius than toe side, or front, sidecut radius. This follows because greater angulation is possible on the toe side, or front, of the snowboard—that is, snowboarders can bend their knees and lean forward into a turn, angling the toe side edge of the board into the snow, to a much greater degree than they can lean backward—so the actual turning radii for the two edges of the snowboard become comparable. Furthermore, snowboards may be asymmetrical, with their heel edges shorter and displaced toward the tail relative to their toe side edges. Many snowboards have a shovel at both ends, so they may be ridden equally effectively in either direction. The larger area of the snowboard relative to skis permits better floatation in unconsolidated snow. The snowboard's shorter length leads to much higher edge loading on compacted slopes. The short carving radii of snowboards generate much higher centrifugal loading forces as well.

The dynamic properties of snowboards are, for the most part, very different from skis. The swing weight or moment of inertia about the center of mass of a pair of skis is a little less than for a snowboard, even though the mass of the board is much larger than is the mass for a pair of skis. The snowboarder stands on the board in a manner such that each foot is placed close to the midpoint between the fore and afterbody of the board. Thus the board may be ridden on either the forebody or afterbody portion by loading one or the other foot. The distribution of flexural rigidity for a snowboard is much more that of a uniform beam than is the case with skis.

Torsional rigidity plays little role in the performance of a snowboard except in the case where a board is designed to be soft in the center section. To initiate the transition from one turn to the next, the snowboarder's boots apply opposite torques to the fore and aft sections of the snowboard so that the edging transition is much more rapid than it is on skis, allowing the snowboarder to execute short, quick turns left and right, maneuvering the board in opposing directions while maintaining full control. The snowboarder must be able to twist the board so that, for example, the left front edge bites into the snow to initiate a turn to the left in nearly the same moment that the right rear edge of the board releases, signifying the completion of the previous turn to the right. This process may be observed on the slope by looking down from the lift at a snowboard track. Notice how the snowboard carves a turn on one edge going in one direction, and then almost immediately switches to the other edge when a turn is initiated in the other direction.

Most manufacturers of snowboards have integrated the design of the snowboard, binding, and boot for racing and for freestyle snowboarding. Snowboard bindings should attach both the heel and toe firmly to the board's surface. The bindings do not need safety-release features because both feet are locked to the single board and thus they twist as a unit. Generally, the left foot is forward and the right foot is back; in a "goofy foot" stance, the right foot is forward and the left foot is back. The back foot may be released from the binding for propulsion over a flat surface. The separation of the feet is adjustable, ranging from 35% to 50% of the contact length. The rotation of the boots on the board is also adjustable. Since the boots are not as stiff as ski boots in either forward or backward lean or in lateral motion, the binding has a high back support that extends up the calf. The binding may be adjusted for forward lean. Because of the displacement of the feet, adjustable canting of the boots—that is, applying wedges under the soles of the boots—is necessary to ensure that the base of the foot makes solid, completely flat contact with the surface of the board.

Snowboard boots usually have an inner boot that is laced into the outer shell. The inner boot's lacing may also tie the inner boot down to the outer boot's sole to provide optimum pressure loading to the board. For the same reason, separation between the sole of the boot and the surface of the snowboard is small. The shells of most snowboard boots, except for boots used in racing, are flexible in comparison to alpine ski boots. Snowboard racing boot models, which are designed to be used at high speeds, are more like alpine ski boots: they have the lateral stiffness needed to give the racer the positive control of edge loading necessary for precise carving at high speeds.

CONCLUSION

The technical developments in ski construction and design that we have discussed above are largely responsible for the recent improved performance that all types of skiers at all levels of ability have seen on the slopes. When we look at the carving and control available to alpine skiers, at nordic, freestyle skate skiing techniques, at the increasing prevalence of free-heel, telemark-turning skiers on the groomed slopes of ski areas, and, finally, at exotic mountain locales where adventure skiers weave through deep powder, we observe a variety of equipment that has greatly increased the variety of experiences available to skiers. It is definitely truer today than it has been in the past that a skier's ultimate performance depends on the skier's equipment and how it is matched to the skier's ability and to the type of skiing the skier pursues.

The dynamic properties of the equipment that are given in Tables 3.4 and 3.5 demonstrate how much range there is across even those abbreviated samples of ski and snowboard equipment. Skiers and snowboarders should match the properties presented there with the maneuvers they wish to attempt and then ask the technical representatives and salespersons specific questions about the contact length, the sidecut, and the sidecut radius of a particular ski of interest. For example, inquire about the sidecut radius and the afterbody bending stiffness of a ski you are considering buying. A shorter sidecut and a stiffer afterbody should help an aggressive skier predict how well the ski will set and hold its edge when making tight turns on a ski slope covered with hard-packed snow.

Finally, remember that there is only so much that anyone can predict about the performance of a pair of skis simply by noting its technical specifications and sizing the skis up in the shop. Skiers must test skis on the slopes. The manufacturers themselves rely heavily on field tests— conducted by racers and by recreational skiers—to evaluate the performance of their equipment. Skiers, too, should insist on field testing any ski they consider buying. Consider the technical features of the equipment and how they are likely to affect your preferred type of skiing, and then see if those technical features translate into improved performance on the slope. When you know what the ski is supposed to do, one of the performance variables becomes known, and you can then evaluate more precisely that other performance variable—your skiing ability—against how you know the ski is supposed to perform.

REFERENCES

1. For a good discussion of ski manufacture and properties, see B. Glenne, "Mechanics of Skis," in *The Handbook of Snow*, edited by D. M. Gray and D. H. Male (Pergamon, Toronto, 1981); see also *The Ski Handbook* (K2 Corporation, a division of Anthony Industies, Vashon, WA, 1991), which offers information about ski manufacturing that is not otherwise readily available.
2. See R. L. Pizialli and C. D. Mote, Jr., "The Snow Ski as a Dynamic System," J. Dynamic Syst. Meas. Control, Trans. ASME **94**, 133 (1972).
3. For a discussion of the physics of snowboarding, see D. B. Swinson, "Physics and Snowboarding," *Phys. Teacher* **32**, 530–534 (1994).

CHAPTER 4

ALPINE SKIING TECHNIQUES
Gliding, Wedging, and Carving

Skiing can be an extreme sport, requiring precise, split-second decisions that may call for total body movements from the skier. Sometimes these decisions and movements must be made while traveling at speeds that may range up to 100 miles/hour (mph) (147 ft/s). Ski equipment is relatively heavy, and, at the speeds involved, the penalties of a crash or collision can be severe. In short, lots of physics is going on, and we ask the reader to remember that our primary objective is to explain the "why" of skiing to develop a greater sense of what actually happens as we speed down the slope, not to detail "how to" best make that descent. Our goal is to work through the physical "why" of skiing to develop a greater sense of the "how to." In the discussions that follow, we assume that readers have some familiarity with basic ski maneuvers. Any ski maneuver, no matter how complex, can be broken into small pieces that can be treated in simple terms—this is the fundamental principal of analysis, that a whole may be examined and understood by study of its constituent parts. For the most part, basic skiing techniques have been analyzed only after the fact of their performance. Scientific and engineering analysis has only recently been employed to guide the development of new levels of skiing performance.

FUNDAMENTALS OF NEWTONIAN MECHANICS

Understanding the physics of alpine ski techniques requires some discussion of basic Newtonian mechanics. Let us start with the mechanics of motion. Readers not familiar with Newtonian mechanics should concentrate on the general concepts, and, once those general concepts become more familiar, consider the quantitative relationships that describe the cause-and-effect phenomena under discussion [1]. First we need some definitions of the dynamic variables involved, beginning with the concept of *force* itself.

Force

Force is the push or pull of one body upon another. Stretch a rubber band with your fingers; one finger exerts a force in one direction, the other finger exerts an equal force in the opposite direction, and the rubber band stretches in between. The directions of the forces are pointed in space in exactly opposite directions. No force exists without its equal and opposite counterpart. Since forces have a direction as well as a magnitude, they are called vectors. A number alone will not specify the force. We represent vectors by drawing arrows in space where the magnitude of the force is proportional to the length of the arrow. The placement of the arrow denotes the point at which the force is applied; thus if the arrow, which denotes the vector force, is moved parallel to itself, it is still the same arrow or vector force, but that force is applied at a different point.

More than one vector force may be applied at a given point. See the skier represented in Fig. 4.1. As a convention, all vector quantities are shown in text as bold characters; without the emboldening, the same character represents magnitude only.

The illustration is a force diagram of a skier in a tucked position skiing down the fall line. The vector forces represented include $\mathbf{W}$, the gravitational weight, resolved into two component forces: $\mathbf{F}_S$, the force parallel to the slope, and $\mathbf{F}_N$, the force normal, or perpendicular, to the slope. The geometric construction used to decompose $\mathbf{W}$ to form the component forces is shown. The force of the skier acting perpendicularly to the snow is the component of his weight, $\mathbf{F}_N$; the two individual forces that compose $\mathbf{W}$ will be used to analyze the skier's motion when all the other forces are taken into account. The other forces that will influence the skier's motion are $\mathbf{F}_D$, the aerodynamic drag; $\mathbf{F}_{\text{plow}}$, the snow plowing force; $\mathbf{F}_f$, a sliding friction force; $\mathbf{F}_L$, an aerodynamic lift force; and $\mathbf{F}_{\text{reac}}$, the sum of the distributed snow pressure forces. The forces illustrated in Fig. 4.1 are shown at the approximate points where they act. We will refer to these

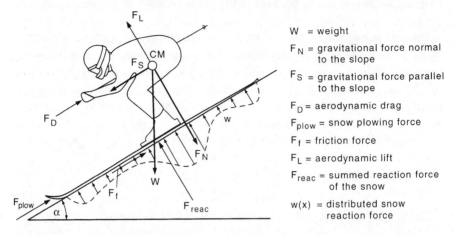

W = weight

F_N = gravitational force normal to the slope

F_S = gravitational force parallel to the slope

F_D = aerodynamic drag

F_{plow} = snow plowing force

F_f = friction force

F_L = aerodynamic lift

F_{reac} = summed reaction force of the snow

$w(x)$ = distributed snow reaction force

FIGURE 4.1. A skier descends directly down the fall line. The various forces that act upon the skier in this situation are shown.

forces using this notation in several different contexts in the discussions that follow.

Displacement

Displacement is the geometric position vector that locates an object at a point relative to some origin. In this case, displacement refers to the position of the skier on the slope relative to the skier's starting point, which we may assume was the top of the ski run. Displacement is a vector because the direction the skier travels must be specified as well as the magnitude to give the distance.

Velocity

Velocity is the change in displacement d per unit of time. Velocity is a vector, and it is given as $\mathbf{v} = (\mathbf{d}_2 - \mathbf{d}_1)/(t_2 - t_1)$, where the displacements are the values given at the times t_1 and t_2. The change in displacement is a displacement vector, so $\mathbf{v}$ is a vector. The displacement vector, $\mathbf{d}_2 - \mathbf{d}_1$, noted in the equational representation of velocity may have any direction.

Acceleration

Acceleration is a vector that describes the change in velocity per unit of time, and it is given as $\mathbf{a} = (\mathbf{v}_2 - \mathbf{v}_1)/(t_2 - t_1)$. From Newton's equation of motion, acceleration of a body is produced by a force acting on the body. If no forces act, then $\mathbf{a}=0$ or $\mathbf{v}_2=\mathbf{v}_1$, so the body stays in a constant state of motion. The velocity does not change in magnitude or direction. The constant of proportionality between the force acting and the acceleration is a quantity called the mass, or M, of the body.

Mass

Mass is an intrinsic property of matter that, through Newton's equation of motion, relates the magnitude of the applied force to the acceleration. Thus $\mathbf{F}=M\mathbf{a}$. Because metric or international units are often used interchangeably with English units in discussions of physics, we should say something about their relationship to each other. In the *Système International* or SI units, mass is measured in kilograms and force is given in newtons when the acceleration is given in meters per second squared. In the English system, mass is measured in pounds; common usage has assigned the weight or the force of gravity on a mass of one pound as the unit of the "pound of force," W. Thus the mass of a body equals its weight in pounds divided by the acceleration of gravity at the surface of the earth, or 32.17 ft/s^2, and this value is given in units called slugs. So Newton's equation becomes $\mathbf{F}=(W/32.17)\mathbf{a}$, where 32.17 ft/s^2 is the acceleration of gravity at the surface of the earth. When M, the mass, is given in pounds of weight, in the expression $\mathbf{F}=M\mathbf{a}$, one must divide the relation by 32.17 to have the force $\mathbf{F}$ given in units of pounds of force. Inertia is another name used in place of the mass. The gravitational force on one pound of mass becomes 32.17 units or poundals, which in the English system is the unit of force equivalent to the metric unit, the newton. In the discussions that follow, we speak of force in units of pounds or newtons.

Angular motion

Skiing involves rotational, or angular, motion, which may be described using the terms *angular velocity* and *angular acceleration*. Angular velocity is the rate of rotation, and it is usually measured in radians per second. There are 2π radians in one revolution; angular velocities in terms of revolutions per second are converted to radians per second by multiplying by 2π. The direction of angular rotation is given by the clockwise motion of a right-handed screw as it is rotated. If ω is the angular velocity, the angular acceleration Ω is defined by $\Omega=(\omega_2-\omega_1)/(t_2-t_1)$. The reader may recognize that this expression is similar to $\mathbf{a}=(\mathbf{v}_2-\mathbf{v}_1)/(t_2-t_1)$, the expression given above in the definition of acceleration. The standard unit of angular acceleration is radians per second squared.

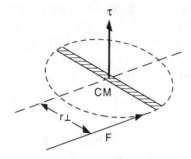

FIGURE 4.2. A force applied to a body produces the torque, τ. The perpendicular lever arm distance from the CM, $r_\perp$, produces the torque $\tau=r_\perp F$ in the direction shown.

Torque

Torque, given as the vector τ, is the effect that causes the rotation of a body. Consider the stick illustrated in Fig. 4.2. If a force **F** is applied at the balance point, given in the figure as the center of mass (CM), the stick moves without rotation. If the same force is applied at any other point, the stick rotates in a manner proportional to $r_\perp$, the perpendicular lever arm (defined as the distance from the direction of the force to the balance point) times the force. Thus torque is the product of the force and the length of the perpendicular lever arm, $r_\perp$, and it is given as $\tau=r_\perp\mathbf{F}$. The unit of torque is either the "pound foot" or the "Newton meter."

The analogous equation of motion for rotation of a rigid body is $\tau=I\Omega$, where I is the moment of inertia. The moment of inertia measures the inertia of a body for angular motion, just as mass measures the inertia of a body for linear motion. From the dimensions in the equation of motion for rotation, one finds the units of I to be pound feet squared or kilogram meters squared. In the sales literature for skis, the moment of inertia is referred to as the swing weight, because swing weight determines the angular acceleration a given torque may produce in a ski.

Momentum

In qualitative terms, the amount of motion may be related to a quantity called the momentum. Quantitatively, momentum **P** is defined as $\mathbf{P}=M\mathbf{v}$. Momentum is a vector quantity that has the same direction as **v** and the dimension of mass times velocity. Newton's equation of motion is more precisely stated as $\mathbf{F}=(\mathbf{P}_2-\mathbf{P}_1)/(t_2-t_1)$. Because **P** measures the state of motion of a body, Newton's first law of motion states that if no force acts on a body, the state of motion of that body remains unchanged.

Remember that force in this context refers to the vector sum of all the forces acting on the body at a given time. Earlier we saw that for every force there is an equal and opposite reaction; that is, on every body the sum of all forces is zero: $\Sigma \mathbf{F}_i = \mathbf{0}$. At least one of the forces in the sum may be an inertial force, that is, the reaction of the mass of the body to the change in motion caused by all of the other forces acting. If all of the applied forces acting on the body sum to zero, there is no change in the state of motion, and thus there is no inertial force. But if the applied forces do not sum to zero, there is a change in the state of motion, and hence there is an inertial force.

The change in momentum defines a direction; the inertial force is assigned a magnitude equal to the vector sum of all the other forces, but it is directed opposite to that sum and thus is opposite to the change in momentum. Thus statics and dynamics problems are treated in the same manner. The concept and application of inertial forces are very important in skiing; it is crucial to understanding ski maneuvers because the skier's momentum changes continuously in magnitude and direction as the skier makes turns and skis down the slope.

Summary of the Dynamics of Motion

Before we move on to consider some other physical properties and eventually consider skiing on the slopes, let us summarize Newton's three laws of motion as they apply to linear and rotational motion viewed from an inertial frame, that is, from a frame of reference that does not itself undergo accelerations of any sort. The three laws are stated as follows:

1. Bodies in motion that are subject to no net forces remain in a constant state of motion.

2. If a net force acts on a body in motion, the body's state of motion must change.

3. For every force acting on a body, there is an equal and opposite force or reaction.

The state of motion of a body is defined by its momentum: for linear motion, the expression is $M\mathbf{v}$, or mass multiplied by velocity; for rotational motion, a similar expression defines angular momentum.

The first law tells us that a moving body continues to move in a straight line unless a force acts to change the direction of the motion (as when a skier moves into a turn) or to accelerate the moving body in the same direction (as when a skier picks up speed by skiing down the fall line). Similar examples may be seen for rotational motion.

The second law leads to the quantitative equations that describe the motion of any mass body. In its simplest form, the second law says that $F = Ma$; that is, the change in the momentum of a body is given by Ma, mass multiplied by acceleration, whether the direction of motion changes or not. This statement has led to more discussion among skiing and science professionals regarding the physics of skiing than almost any other single idea. The term Ma must be a force because it equals the force F, but it is a force that has the direction opposite to F. Thus we may say $F + F_I = 0$; that is, a force F plus the inertial force F_I equals zero. So the inertial force F_I equals the negative of mass times acceleration, or $-Ma$. Note that this exposition is consistent with the third law: the inertial force is the equal and opposite counterpart of the applied force acting.

The third law tells us, among other things, that a skier standing perfectly still on a slope has the force of gravity pulling that skier into the snow. The skier may sink downward into the snow until the snow under the skis compacts and generates an equal and opposite upward force on the bottom of the skis, counteracting the force of gravity, and the skier stands still.

Work

When a force is applied to a body and this action results in motion in the direction of the force, *work* will be done by the force. If no displacement of the body occurs in the direction of motion, or if the displacement is at right angles to the force, no work is done. Work is measured in units of force times the displacement in the direction of the force, and it is given in units of foot pounds or newton meters. To avoid confusion given the similarity in nomenclature, units of work are given in foot pounds, and units of torque are given in pound feet. Thus a 150-lb skier lifted vertically by a ski lift 1000 ft up a mountain has 150 000 ft lb of work done on him. Any horizontal motion results in no additional work. Similarly, if it requires 3 lb of force to pull our 150-lb skier over level ground, moving that skier 1000 ft would require 3000 ft lb of work.

Power

Power is the time rate of doing work. Suppose a 150-lb person climbed 1000 ft in 60 min, or 3600 s. The average power expended to move the body up that incline would be 150 times 1000/3600, which equals 41.7 (ft lb)/s. Since 1 horsepower (hp) is 33 000 ft lb/min, or 33 000/60 = 550 (ft lb)/s, this person worked at the rate of 41.7/550, which, expressed in terms of power, yields 0.076 hp. This power value does not accurately represent the total working rate of the person because additional work goes into just walking on a level surface and the ongoing operation of bodily functions. The total work rate of our hypothetical person, including all bodily functions, is from 1 to 1.5 hp, with a maximum power rating for climbing uphill of 2. 2 hp.

Energy

Work and energy are identical. The work done to lift a skier up a hill that results in a change in elevation increases that person's potential energy by an amount equal to the work performed. Potential energy is energy stored for future use. It may be transferred and stored by an appropriate mechanism, such as the energy of compression that is stored in a spring or in the pressurized gas stored in a vessel. Consider the example of a skier who has been lifted to the top of the right-hand hill at an ideal ski area that sits in the bottom of a valley with equally high hills on either side, left and right. If the skier chooses to ski back down the hill (a natural choice), the skier converts the potential gravitational energy that she stores in her situation at the top of the hill (work has been expended to change her elevation by lifting her to the top of the hill) into energy of motion, called kinetic energy. She points her skis down the fall line, and down the hill she goes.

While potential energy may exist in many forms—gravitational as in the example of our ideal skier and ski area discussed above, internal deformation of a rigid body as in a spring or bent ski, chemical as given by the charge in a storage battery or in the pasta we eat for supper—there is just one form of kinetic or motional energy. The quantity of kinetic energy, T, depends only on the mass and velocity of motion. From Newton's equation of motion, a force ($\mathbf{F}$) acting in the direction of the momentum ($\mathbf{P}$) of a body changes the magnitude of the momentum according to the relation $\mathbf{F}\Delta t = \Delta \mathbf{P}$, where Δt is the time interval over which the force acts and $\Delta \mathbf{P}$ is the change in momentum. Multiply this relation by the velocity v to get $Fv\Delta t = F\Delta d = Mv\Delta v$, since $P = Mv$. Given that $F\Delta d = \Delta T$, upon summing both sides of the relation, $\Delta T = Mv\Delta v$ for the case in which the mass starts at rest, one gets the following relation for the kinetic energy T of a mass body

$$T = \tfrac{1}{2}Mv^2. \tag{4.1}$$

By now the skier at our ideal ski area has descended the right-hand hill. Assuming an ideal situation in which no forces act to cause gains or losses of energy in her descent, the skier reaches the bottom of the hill with an amount of energy equal to the potential energy that was stored in her trip to the top of the right-hand hill, and she will now slide up the hill on the left-hand side of our ideal ski area, arriving at the top of that hill and coming to a stop. At that point the skier could turn around and do it all over again in the other direction, skiing up and down and back and forth perpetually. Clearly this cannot happen in reality. As we shall see, when we ski real slopes there are many forces that dissipate the gravitational energy that we convert to kinetic energy in our descent. But the notion of skiing up a hill as fast as we ski down it is intriguing, and it demonstrates how energy is conserved and converted.

This example brings us to an important principle that provides insight into many kinds of processes: the conservation principle, which states that the work done on a system must appear in the system as an increase in the sum of potential and kinetic energy. In skiing, the primary source of potential energy (PE) is gravitational energy, which is given by the relation $PE = Mgh$, where M is mass, g is the acceleration of gravity, and h is the vertical elevation above some reference elevation, say the bottom of the slope. If no work is done on the system, the sum of the potential energy, $PE = Mgh$, and the kinetic energy, $T = \frac{1}{2}mv^2$, remains constant. The conservation of energy relation for this case states

$$PE + T = \text{constant},$$

thus

$$Mgh + \tfrac{1}{2}Mv^2 = \text{constant.} \tag{4.2}$$

In the idealized example we considered above, no work was done on the system by outside forces. If an outside force such as the wind had blown on our skier, the total energy in the system would have increased or decreased, depending on the direction of the wind, and the skier would have either gone past or failed to reach the top of the left-hand hill accordingly. Energy may be exchanged between potential and kinetic forms, but the sum of the energy in a system changes only if work is done on the system. Remember that this statement is not precisely correct for an actual skier because the human body has a reservoir of potential chemical energy stored within it in the forms of glycogen and fat. Also, when the skier does physical work—flexes her muscles to edge her skis and so on—the mechanical energy may be changed. Nevertheless, the conservation principle allows us to draw conclusions about the energy relationships that exist in a system even if we cannot follow in exact, minute detail everything that happens in that system.

DYNAMICS OF STRAIGHT GLIDING

To begin our discussion of skiing, let us first work out the mechanics of straight gliding on an inclined snow slope [2]. Figure 4.3 shows the geometry associated with and the forces acting upon a skier following the fall line as he glides down a slope. The skier's center of mass (CM), or where the total mass of his body may be located to fix the force of gravity or weight, is set just at the skier's hips. The weight **W** may be resolved into two forces, F_S parallel to the slope, and F_N normal to the slope. The reaction of the snow to the ski is shown as the force F_{reac}, which is the sum of the distributed forces due to the snow pressure on the bottom of the ski. The skier gliding down the slope has a snow friction force F_f and a wind resistance or aerodynamic drag force F_D, which are directed up the slope. An aerodynamic lift force F_L is also present. This case represents a skier who achieves terminal velocity; there is no acceleration and thus there is no inertial force, so all the forces as well as the torques shown must sum to zero.

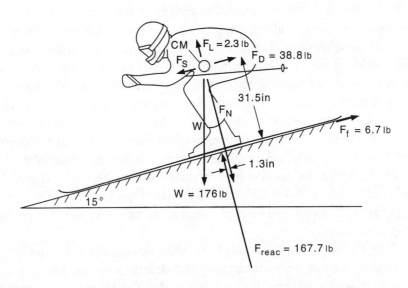

FIGURE 4.3. A skier descends directly down the fall line. Some representative forces are shown in pounds. The skier's mass W equals 176 lb. The skier's velocity of 63 mph, or 92 ft/s, is constant or terminal. F_D, the aerodynamic drag, is 38.8 lb, snow friction F_f is 6.7 lb, and the aerodynamic lift F_L is 2.3 lb. All of the forces except the snow friction act at the CM, so the snow reaction force F_{reac} must act in front of the CM, as shown (Perla and Glenne, 1981).

The forces are shown at the points on which they act. The gravitational force acts at the center of mass. The aerodynamic drag ($\mathbf{F}_D$) acts at the effective center of the frontal area, in line with the center of mass in this case; the snow friction force ($\mathbf{F}_f$) acts along the contact area where the ski meets the slope. The forces shown are for a limiting case of constant velocity for a 176-lb skier moving at 63 mph, or 91.8 ft/s, with a friction coefficient of 0.04. The aerodynamic drag and lift are representative of wind-tunnel tests for a skier of this weight traveling at this velocity. For the skier to move down the slope with a stable motion, the sum of all the torques about the center of mass must be zero. While the snow reaction force $\mathbf{F}_{\text{reac}}$ equals the sum of the forces at the ski bottom, the location of the force in the diagram must be such that it produces an equal but opposite torque about the center of mass as the actual distributed forces do. In the illustration, only the friction force ($\mathbf{F}_f$) and the snow reaction ($\mathbf{F}_{\text{reac}}$) do not act through the center of mass. Hence the snow reaction force ($\mathbf{F}_{\text{reac}}$) must act in front of the center of mass at the distance shown in order to balance the torque of the friction force. The friction force tends to pitch the skier forward, so the forebody of the ski must develop enough additional force to stabilize the body.

Recall Fig. 3.8 from the preceding chapter in which some characteristic pressure distributions along the bottoms of several skis were illustrated. Most of those pressure distributions were symmetrical about the boot position, but some were not. Furthermore, the skier's center of mass is not fixed directly above the boot, so there may be a resultant nonzero torque. The term *resultant* refers to the sum of a number of vectors. Under these resultant nonzero torque conditions, the skier's legs and ankles will be uncomfortable, so the skier, by means of biomechanical feedback control, bends his upper body, hip joints, and knees, moving the center of mass to make the torque on the system (the skier and the skis) zero, if possible. Because alpine ski boots are fixed to the ski at the toe and the heel, such adjustments in torque can be transmitted to the skis to change the pressure load distribution.

Using the forces we discussed above when we considered Newton's laws of motion, we can write an equation that describes the motion of a skier moving down the fall line. Recall that $F = Ma$. The force F that causes the skier's motion is the sum of all of the forces acting parallel to the slope that result from the acceleration of gravity, g, pulling the skier down the slope. The skier's mass, M, may be expressed as his weight W divided by the acceleration of gravity, or W/g. Thus $F = Wa/g$, and the equation becomes

$$\frac{Wa}{g}=F_S-F_f-F_D, \tag{4.3}$$

where F_S is the component of the force of gravity down the fall line, which is impeded by two friction forces that act to slow the skier's descent (and thus are represented as negative values), F_f, is the snow friction force, and F_D is the aerodynamic drag force. The skier's motion down the hill is produced by the force of gravity parallel to the slope minus the snow friction force and the aerodynamic drag force.

The snow friction force exerted on the skis, $\mathbf{F}_f$, is independent of velocity and may be expressed as $\mathbf{F}_f=\mu\mathbf{F}_N$, where μ is the coefficient of friction and $\mathbf{F}_N$ is the normal reaction force, although this relation is not precise. Thus one may say that "friction is a traction that is a fraction of the normal reaction." The aerodynamic drag ($\mathbf{F}_D$) depends on the square of the velocity, except at very low speeds where it depends on velocity only. Aerodynamic drag also depends on the magnitude of the frontal area of the body, A, the density of the air, ρ, and a drag coefficient, C_D. The equational expression for aerodynamic drag follows:

$$F_D=\frac{C_D A \rho}{2}\, v^2. \tag{4.4}$$

The drag coefficient (C_D) depends on the velocity and the aerodynamic shape of the skier. Additional drag forces, the snow compaction force $\mathbf{F}_{comp}$ and the plowing force $\mathbf{F}_{plow}$, are encountered when skiing on new or unpacked snow. Note that acceleration is independent of the weight of the skier if we neglect to consider the drag forces generated by wind resistance, snow compaction, and plowing.

Using the equation of motion for a skier moving down the fall line, we can describe the ultimate speed of a downhill racer, that is, the speed at which the skier's acceleration a equals zero. When we rearrange Eq. (4.4) by inserting the expressions for F_S, F_f, and F_D, adding A for the area of the skier's body facing the direction of motion, and setting the acceleration a at zero ($a=0$), the expression becomes:

$$F_S-F_f=F_D, \quad F_f=\mu F_N=\mu W \cos\alpha, \quad F_S=W \sin\alpha, \quad F_D=\frac{C_D A \rho v^2}{2}$$

hence $\hspace{10cm}$ (4.5)

$$\sin\alpha-\mu\,\cos\alpha=\frac{C_D A \rho v^2}{2W}.$$

Figure 4.4 shows a graphical analysis of this equation. Once μ is specified, the appropriate curve versus the slope angle α is determined. Values for C_D, W, A, and ρ must be inserted to calculate the coefficient for the curves plotted against v^2. The air drag factor D becomes $D = C_D A \rho / 2gM$, with $W = gM$ in newtons. Follow the dotted line from the slope angle and move up to the curve for μ, then over to the curve on the right half of the graph for the appropriate air drag factor, and then down to determine the velocity v.

To calculate an actual example of velocity using Eq. (4.4), let us insert some typical values for a skier skiing downhill:

$$C_D = 0.5, \quad A = 0.4 \text{ m}^2, \quad \rho = 1.2 \text{ kg/m}^3,$$

$$\mu = 0.04, \quad \alpha = 15°, \quad M = 80 \text{ kg}.$$

When we solve the equation for v, we find that $v = 38$ m/s.

We can make several observations from our use of relation (4.4) to describe a downhill skier's descent of a slope. First, the aerodynamic factors A and C_D have the most influence on a skier's ultimate speed. The frontal area A changes as the skier leans forward to counter the moment produced by the aerodynamic drag; it changes most when the skier goes into a tucked position. C_D varies from a value as high as 1.2 for a slow skier (treated as a cylinder moving in a fluid) to a value of 0.2 for a fast skier in

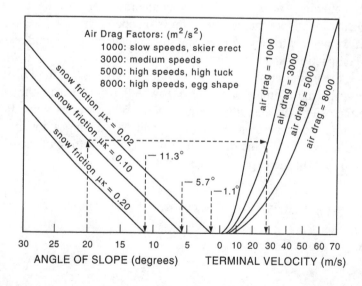

FIGURE 4.4. Limiting velocity nomograph of Eq. (4.5) (Perla and Glenne, 1981).

a tucked position (treated as a sphere in the same moving fluid). The density of air decreases with elevation, but it increases with the lower temperatures encountered at higher elevations; the change may be as much as 40%.

Aerodynamic drag depends on the square of the velocity, so changes in the parameters of the downhill skier's schuss have a smaller influence on the terminal velocity as the speed increases. Thus a fast skier may control speed by varying the aerodynamic drag through changes in A or C_D without exerting much physical effort. Ski racers are admonished never to get airborne. Why? From the aerodynamics of flight it can be shown that aerodynamic lift as a force transverse to the direction of motion has an associated drag that varies with the lift as v^2. Even on the ground, the lift may be downward. Whatever the direction of the lift, the associated drag adds to the intrinsic drag on the body to make the skier's ultimate speed decrease.

In reality, most downhill skiing is not done in a manner consistent with the model we are using here, a free, sliding body on a slope. Skiers may exert poling or skating forces, setting into play plowing and skidding forces as they move through series of turns down the slope. When we add F_{pl}, the poling force, to our equation for the motion of a skier moving down the fall line, we get

$$\frac{Wa}{g} = F_S - F_f - F_D + F_{pl}. \tag{4.6}$$

The poling force F_{pl} is needed to accelerate and break out of the domain where the static coefficient of friction (μ) is large and move into the range where μ is low, the dynamic or sliding domain.

The slower skier must control speed by plowing or skidding, which requires more muscular exertion. The friction coefficient μ may range from 0.02 to 0.2, but the usual range will be from 0.02 to 0.05. Sliding friction is a very complicated process. The coefficient may have a component that depends on the velocity v. Plowing forces, which depend on pushing or cutting snow in the path of the skis, depend on v^2, so they change the ultimate speed markedly.

The braking force produced by plowing or skidding on the ski edge can be estimated using the conservation of energy principle. The work done by the braking force F to stop the skier in a distance d down the slope is Fd, which must equal the decrease in kinetic energy, so $Fd = Mv^2/2$. To put the force in units of weight, divide by the acceleration of gravity g to obtain $F = Mv^2/2gd$. Now, add the component of weight parallel to the slope, F_S, and the total braking force F becomes

$$F = \frac{Mv^2}{2gd} + F_S. \qquad (4.7)$$

For example, a 160-lb skier stopping in 15 ft from a speed of 30 mph, or 44 ft/s, on a 15° slope exerts a braking force that becomes 364 lb. Unless the skier is fully in control of this maneuver, such a force may cause serious injury.

MECHANICS OF THE TRAVERSE

For the most part, skiing is done by making successive traverses of a slope and linking the traverses with turns. Skiers should understand the geometry of the traverse because the angle that forms between the plane of the ski and the slope determines how the ski carves the snow. Figure 4.5(a) shows a slope tilted at an angle α with a skier traversing downslope at an angle β, relative to horizontal. A diagram of the corresponding forces that relate to the skier's downslope traverse is presented in Fig. 4.5(b). Here the gravitational force **W** is resolved into two components: $\mathbf{F}_N$, the force normal to the slope plane, and $\mathbf{F}_S$, the force down the fall line, or $W \sin \alpha$, where α is the angle of the slope. The gravitational force down the fall line, $\mathbf{F}_S$ or $W \sin \alpha$, is further resolved into $\mathbf{F}_p$, the force parallel to the ski track, and $\mathbf{F}_{\text{lat}}$, the force perpendicular to the track direction. The force along the direction of motion causes the acceleration and the inertial force $\mathbf{F}_I$, and hence the motion. To maintain proper balance, the skier adjusts his body position laterally, so the remaining component of weight, $\mathbf{F}_{\text{load}}$, is perpen-

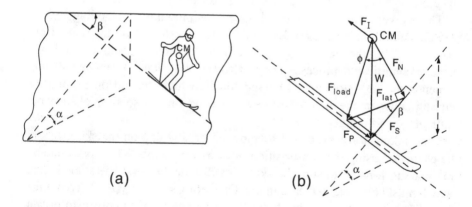

(a)   (b)

FIGURE 4.5. (a) Skier on a downslope traverse. (b) The resolution of forces to determine F_{load}, F_{lat}, F_p, and the tilt angle ϕ.

dicular to the plane of the ski. Seen from the perspective of the balance of gravitational forces, a skier traversing a slope does not lean sideways: the skier's weight is directly over the ski.

With this definition for the geometry of the plane of the ski, we can derive expressions for several of the forces involved and one of the angles: the force in the direction that the skier is moving, $\mathbf{F}_p$ [Eq. (4.8a)]; the force perpendicular to the track direction, $\mathbf{F}_{lat}$ [Eq. (4.8b)]; the force perpendicular to the ski, $\mathbf{F}_{load}$ [Eq. (4.8c)]; and the skier's tilt angle ϕ that forms between the line that indicates the direction of $\mathbf{F}_{load}$ and the plane of the slope [Eq. (4.8d)].

$$\mathbf{F}_p = W \sin \alpha \sin \beta, \qquad (4.8a)$$

$$\mathbf{F}_{lat} = W \sin \alpha \cos \beta, \qquad (4.8b)$$

$$\mathbf{F}_{load} = W(\cos^2 \alpha + \sin^2 \alpha \cos^2 \beta)^{1/2}, \qquad (4.8c)$$

$$\tan \phi = \tan \alpha \cos \beta. \qquad (4.8d)$$

In these expressions, $\mathbf{F}_p$, the force parallel to the ski, causes acceleration and is equal and opposite to the inertial force, $\mathbf{F}_I$. The force perpendicular to the track direction, $\mathbf{F}_{lat}$, if it is great enough, may cause the ski to skid out of its track. The total force perpendicular to the ski, $\mathbf{F}_{load}$, bends the ski and forces it into the snow. Note that if $\beta = 0°$, $\phi = \alpha$; or, when a skier stands with his skis pointed horizontally across the slope, $\mathbf{F}_p = 0$ and $\mathbf{F}_{reac} = W$. When the skier makes a run directly down the fall line, $\beta = 90°$ and $\phi = 0$. These equations are discussed in grater detail in Technote 3, p. 203, "The Loads on a Running Ski."

EDGING AND ANGULATION

Except when skiing straight down the fall line, the ski is always set on edge into the snow so that it makes an angle, Φ, with the slope. Because the edges on a modern ski are not straight lines—the ski has a narrower waist, as we have seen—the contact line the ski makes with the slope surface generally approximates the arc of a circle. Thus, when the ski runs forward on a traverse, it tends to curve uphill. To counter this effect, skiers may sit back on the tails of their skis so that a short length of the edge on the afterbody of the ski is the only part of the ski's edge in contact with the snow. The skier may also roll the skis so that the angle Φ that forms between the ski and the snow surface decreases, and the skis then tend to skid slightly sideways. Using this maneuver, commonly called *sideslipping*, a skier may skid sideways down a slope without making any motion at all in the direction in which the ski tips point. The processes of edging and wedging, whereby the skis do not move in the direction they are pointed but sideslip over the slope, are fundamental to both the steered turn and to braking to control speed.

Once a skier advances beyond using the wedge and skid style of turning to making carved turns, the maneuvers needed to control the edge angle during the turn are called *angulation*. A skier standing on skis at rest and flat footed may, by rolling the ankles or torso, shift his weight so that it applies to the inside or outside edges of his skis, transmitting the force through the soles of the boots. Ski boots generally do not allow for much ankle rotation, so the legs and upper body must flex to roll, or angulate, the soles of the boots, setting the edges of the skis into the snow. The photographs in Fig. 4.6 illustrate two methods of angulation. The pointers on the tips of the skis show their edge angle, and the stripes on the skier's trousers show the effect of the angulation maneuver in the line from the hip joints to the knees and to the boots. The photograph on the left shows the effect of rolling the flexed knees; if the legs are straight, this angulation maneuver cannot be done. The skier's upper body rotates in the same direction as his

(a)　　　　　　　　　　　　　　　　　　　　　　　　　　(b)

FIGURE 4.6. Dave Lind demonstrates static body angulation. (a) illustrates angulation produced by body rotation with the knees bent; (b) illustrates the angulation produced by the reverse-shoulder or counterrotation maneuver.

knees; this angulation maneuver is called *rotation*. By rotating the upper body from the hips up to the shoulders to the right, the skier makes his knees roll, which edges the skis onto their right edges. This maneuver requires that the skier bend his knees and maintain the angle that forms between his lower and upper leg, and then rotate his entire upper body about a vertical axis, which causes the soles of his boots to edge the skis. If the turn is to the left, the body rotates to the left; if the turn is to the right, the body rotates to the right. There is a definite limit to the edge angle Φ that can be generated using the rotation maneuver.

The preferred angulation maneuver is illustrated in Fig. 4.6(b) in which the skier's shoulders and torso are rotated in a direction opposite to his knees, causing the skier's torso to bend sideways at the hips. The skier's knees do not rotate, and the skier's hips back in toward the slope. The sideways roll of the soles of the feet produced by this "counterrotation" or "reverse-shoulder" angulation maneuver does not depend on the knee bend, it depends on the hip bend. If the hip goes to the right, the boot sole rolls to the right edge; if the hip goes to the left, the boot sole rolls to the left edge. Thus, when the turn is to the right, the body and shoulders turn to the left, and vice versa, which explains why this angulation maneuver has been called counterrotation.

Note that the edge angle of the skis that can be achieved by using counter-rotation is much larger than the edge angle produced by the rotation maneuver. Note also that throughout these edge angle maneuvers, in spite of the stiff outer shell of the ski boot, the skier can roll his foot at the ankle inside the boot to make subtle changes in the effective angulation of the ski. In this manner, the skier can hold or release the edge for carving or skidding as needed. During a traverse of the slope, the edge angle of the two skis may not be the same, nor will the loading be equal; thus some skidding and edge release must occur for the skier to ski a straight track.

Readers can get a feel for the body configurations involved in these angulation maneuvers by trying out both rotation and counterrotation while they wear their ski boots and stand on a flat floor (there is no need to be on a ski slope to feel the weighting on the edges of the soles of the boots). Note that neither rotation maneuver can be accomplished very well when the upper body is bent forward at the waist, as is so frequently the case with beginning to intermediate skiers. In the early days of ski instruction, the most common admonition was "Bend your knees." In part because of our better understanding of angulation, that advice has changed to "Stick your chest out" or "Keep your body upright." For the beginning to intermediate skier, angulation is primarily a means of controlling speed and of negotiating a steep pitch by using sideslip through alternate edging and edge release. For more advanced skiers, angulation is the key to the carved turn.

MECHANICS OF WEDGING AND THE WEDGE TURN

Beginning skiers often feel helpless and terrified on the slope until they learn a reliable technique that will allow them to control their speed and stop. So, soon after learning the simplest maneuvers for walking on skis over level or nearly level surfaces, beginning skiers usually learn to negotiate a gentle slope by letting their skis slide flat on the slope with the tips together and the tails apart. In this triangular attitude, the skis form a V shape and the skier performs a braking and turning maneuver that has been variously called the wedge, the pie, and the snowplow. The wedge stance allows the skier to slow the rate of descent because when the skis are flat on the snow their lateral skidding generates a plowing force on the outside edges of the skis, which balances the gravity force down the fall line as well as the separation imposed by the leg thrust. In the wedge stance, the legs naturally load the skis on their inside edges. By bending the legs at the knees and rotating the knees together, the load concentrates on the inside edges of the skis, creating a large edging angle Φ, which in turn generates a large plowing force that works to slow or stop the downslope motion.

As the slope angle α increases, the wedge angle that forms between the two skis and the edge angle must also increase for the plowing force to be sufficient to slow the skier's descent. On a steeply inclined slope it may become impossible to open the wedge angle enough with the snowplow maneuver to generate the edge-cutting force needed to slow or stop the skier, and an alternative braking maneuver must be used. This maneuver is the parallel skid, or skater's stop, in which the skis are set on edge horizontally across the slope. Using added body angulation to increase the edging angle and maximize the edge-cutting force of the ski, the skier forces the uphill edges of the skis into the slope in a manner analogous to an ice skater's rapid stop. The force on the ski perpendicular to the slope and the edging angle determine the cutting action of the edge. The stance used in the parallel-skid braking maneuver—with the legs, knees, and feet together, the skis parallel to each other and across the fall line—is more comfortable and safer than the wedge stance, and the edge braking action can be readily controlled on a variety of surface hardness conditions. For these reasons, beginning skiers are taught the parallel-skid braking maneuver as soon as possible. The confidence that beginning skiers gain from knowing that they can control their speed encourages them to let their skis run parallel downhill, so that they can go on to learn more advanced turning maneuvers.

Once the beginning skier is willing to let the ski wedge point down the fall line, the ski configuration and weight distribution created by the snow-plow position generates the wedge turn. The maneuvers involved in a typical sequence of wedge turns are shown in Fig. 4.7. By leaning to place his weight onto the right or the left ski, the skier learns that he can turn to the left or to the right, respectively. The weighted ski becomes the outside ski in the turn. Let us look a little more closely at the forces involved in making these wedge turns.

In a steady state, that is, when sliding at a constant velocity, the total force on each ski is transmitted up the leg to the center of mass of the body, and the vector sum is about equal to the downslope force of gravity, or weight [see Fig. 4.8(a)]. The force diagrams in Figs. 4.8(b) and 4.8(c) show the projections of all of the forces on the plane of the slope. The direction of the force on each leg is nearly perpendicular to the direction each ski is pointed. The ski slides easily along its length, but it does not slip sideways. For motion straight down the fall line, the snow reaction forces on the right and left legs are equal and their sum passes through the center of mass [see Fig. 4.8(b)]. If the skier shifts his weight toward the right ski, the forces may change in magnitude, but not much in direction. Now the component

FIGURE 4.7. Wedging to initiate a turn with a nearly parallel stance at the completion of the turn. [Reprinted with permission from G. Joubert, *Skiing: An Art... a Technique,* Poudre Publishing Co., La Porte, Colorado (1980).]

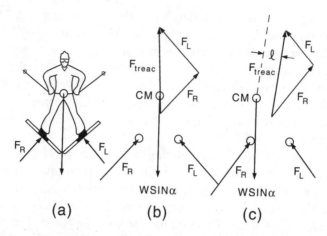

FIGURE 4.8. Force diagrams for constant velocity wedging downhill. In (a), the forces F_R and F_L and reactions on the skis are in the slope plane and normal, or perpendicular, to the skis. (b) shows the resolution of the forces for the linear downslope motion. (c) shows the case in which the CM is displaced to the right to make a left turn. F_{treac} is the reaction of the snow on the skier in the slope plane.

of the gravitational force in the plane, $F_S = W \sin \alpha = \mathbf{F}_{treac}$, where $\mathbf{F}_{treac}$ is the total reaction force in the plane of the slope acting on the skier, is separated by the distance l from the resultant of the two leg forces, so a torque of $lW \sin \alpha$ rotates the skier counterclockwise, and the skier turns to the left. There is also a resultant sideways force acting on the center of mass to move the skier to his left. The torque acting on the skier causes the skis to skid as they rotate about the boot and they slide onto their inside edges. The skier controls the skid by rolling his ankle to decrease the edging action of the skis as needed to make a smooth turn. Thus the transfer of weight from one ski to the other in part controls the rate of the turn and its radius; the control of the edge angle through the knees also controls the relative leg loading and thus the turning.

This points out a concern related to wedge turning. As skiing speeds increase, the chances of catching a ski edge during a wedge turn become greater and greater. If the ski does not skid (rotate while the body rotates), it catches an edge, and a serious rotational torque strain of the knee and leg may result. Torn knee ligaments or even spiral fractures of the lower leg bones can result, even if the ski bindings release. Small children having just learned to control their skis by wedging will often race down the hill in a wedge position. Fortunately, children use very short skis with correspondingly less potential for generating torque in a fall. But even so, it remains true that the physics of wedge braking and wedge turning demonstrate that these maneuvers can be risky when they are attempted at higher speeds and on steeper slopes. These facts point again to the wisdom of having beginning skiers master the parallel-skid braking maneuver as soon as possible. The greater edge angle—and hence greater braking—that is possible with the parallel-skid braking maneuver reduces the risk of injury, as does the fact that the skis and legs are locked together when we parallel skid to a stop. Finally, the relative body and ski position and the movements used in the parallel-skid braking maneuver help introduce some elements of the carved turn.

Wedge braking and turning should be used for low-speed skiing and for skiing in confined tracks, such as down a narrow road or trail or in crud snow with a heavy pack. The wedge stance places considerable strain on the leg muscles, so it should be used only intermittently in recreational skiing. Wedge braking and turning are best for situations in which careful, continuous force control is needed; as when, for example, a ski patroller guides an injured skier on a toboggan down the fall line.

MECHANICS OF THE CARVED TURN

For the most part, ski instruction offered in print by early ski professionals does not explicitly emphasize carving as a turning technique. One of the first comprehensive discussions of the process of carving a turn on skis was published only in the last decade, and our discussion of carving turns draws much of its qualitative analysis from that source [3]. The evolution of modern ski design, in particular the designers' recognition of the roles played by sidecut and the flexural properties of the ski, have had much to do with the evolution of the modern carved turn. Writing at the beginning of this decade, Stenmark makes the point that many skiers do not take advantage of the way modern equipment has been designed to facilitate carving turns [4]. Also, different teaching systems have evolved over the years with different protocols for teaching the accomplished skier to carve turns. The carved turn as it is performed on the slopes today has evolved from all of these systems and protocols. Understanding the underlying mechanics of the carved turn should help skiers recognize and clearly separate the carving processes from the skidding or braking processes involved in the turn. In the carved turn, unlike in the wedge turn, control of speed and control of direction are independent of each other.

Skiing involves making beautiful turns, and turning involves changing the direction of the skier's momentum. That change in momentum requires a force. Consider the case for turning at a constant velocity along a path (see Fig. 4.9). The acceleration, $\mathbf{a}=(\mathbf{v}_2-\mathbf{v}_1)/(t_2-t_1)$, is directed toward the center of curvature; hence it is called centripetal acceleration. The inertial

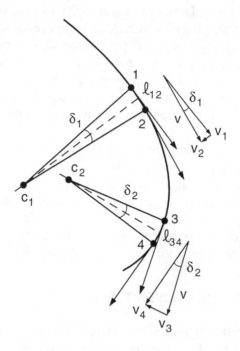

FIGURE 4.9. Geometry of a trajectory for calculating centripetal acceleration during a carved turn.

force, or the centrifugal force, $\mathbf{F}_C = -Ma$, is directed away from the center. The arc length, l_{12} equals $v(t_2 - t_1) = R\delta$, where R is the radius of curvature, and $|\mathbf{v}_2 - \mathbf{v}_1| = v\delta$. Divide the second expression by the first to eliminate δ; then it follows that $a = v^2/R$.

A simple example of the interplay of centripetal and centrifugal force is the motion of a mass swung on the end of a cord. The tension in the cord is the centripetal force, which balances the inertial centrifugal force of the mass and causes the body to swing in a circle. Carving turns on skis involves motion at different velocities in arcs of circles of different radii. Successful skiers develop a fine sense of how to balance their bodies so that the reaction of the snow on the skis just balances the forces of gravity and the centrifugal inertial force, so the skis and the skier turn, making the transition from linear to circular motion.

Usually the motion of carving a turn is not that of a pure circle, but at every point on the curve of the turn, the curvature, R, is defined by the curve. Hence the centrifugal force changes accordingly. Skiers must contend with the conflict of forces that exists between the dynamics of the skier's balance and the geometry of the circular turn. We will simplify the discussion of this phenomenon somewhat by assuming that the skier's velocity remains constant. On the one hand, the dynamics of the skier's balance requires that the resultant of all external forces acting on the skier—the gravitational forces and the snow reaction force that passes through the line of contact of the ski with the snow—must just balance the skier's inertial force. Thus the resultant of all external forces in the plane of the slope acting on the skier must pass through the center of curvature to balance the centifugal inertial force. Another way to say this is that the gravitational and inertial forces acting on the skier at his center of mass must sum to a force that acts precisely through the ski–snow contact line to balance the snow reaction force. On the other hand, the geometry of the edged and flexed ski in contact with the slope leads to the development of the radius of curvature of the contact line. If the ski does not skid laterally, it tracks in the groove generated by the ski's edge, and it *carves* a turn. The circular motion of the turn generates a centrifugal force $\mathbf{F}_C$, which adds or subtracts from the weight (or lateral gravitational force) component, $\mathbf{F}_{lat}$, which is parallel to the slope plane.

These forces and the tracking of a ski are illustrated in Fig. 4.10, which shows the circular track of a ski carving a turn of a constant radius on a fixed incline traversed at a constant speed. The illustration represents a realistic case for a velocity v of 22 ft/s, or 15 mph, and a radius R of 40 ft on a slope with an angle of 30°. The centrifugal inertial force $\mathbf{F}_C$ is constant and is directed outward radially at every point from the circular path of the turn. The lateral gravitational force, given by the relation $F_{lat} = W \sin \alpha \cos \beta$, changes from pointing toward the center of curvature in the uphill quadrants to pointing away from the center of curvature in the downhill quadrants. The sum of the forces shown by the outer of the two dashed curves determines the angle that the skier's body must tilt at for stability. Those tilt angles are shown for selected points along the path of the turn. If the body center of mass (CM) follows the fixed circular path, the ski edge must follow the outer dashed track. Thus, in the uphill quadrants of the turn, the gravitational and centrifugal forces are exerted in opposite directions; in the downhill quadrants of the turn, they are exerted in the same direction.

Refer to Figs. 4.11(a) and 4.11(b) in which the resultant lateral force and total force for a skier edging downhill (a) and uphill (b) are illustrated. The components of weight, $\mathbf{W}$, $\mathbf{F}_N$, and $\mathbf{F}_{lat}$ are shown to calculate the load force $\mathbf{F}_{load}$. In the figure, the skier uses the proper tilt angle, ϕ, to assume the natural, balanced position needed to move through the turn. The centrifugal force $\mathbf{F}_C$, is assumed to be constant. The load force passes through the ski contact line and the body center of mass because the body is properly

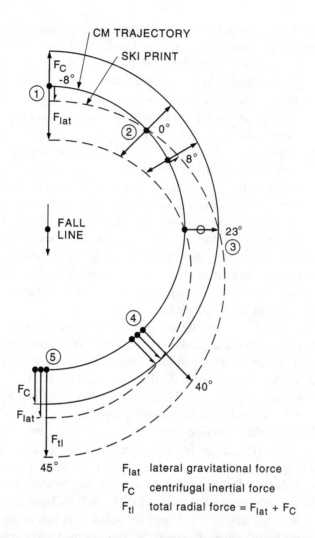

FIGURE 4.10. Forces and ski print for a circular turn of constant radius and velocity. F_{lat} is the lateral gravitational force at each point, F_C is the centrifugal inertial force, and F_{tl} is the total radial force. The CM follows the solid circular path. F_{tl} acts at the CM so the outer dashed curve represents the position of the ski to ensure that the total force acting on the CM passes through the ski; it is the ski print necessary for stable motion. The tilt angles are shown for selected points during the turn.

balanced. With the inclusion of the centrifugal force, the load expression for the carved turn illustrated is complete, unless dynamic leg action or poling forces come into play.

We may now modify Eq. (4.8) from our earlier discussion of traversing directly across a slope to include the centrifugal force term that comes into

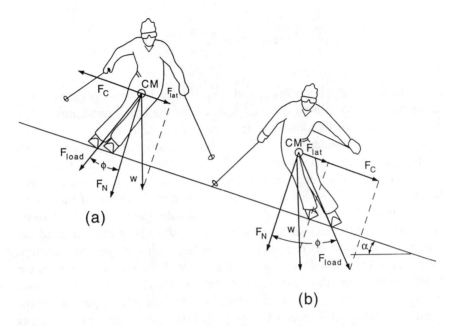

FIGURE 4.11. Lateral force diagrams for a skier in the (a) uphill and (b) downhill quadrants of a carved turn.

play when we carve turns. The total transverse force vector $\mathbf{F}_{tl}$ is given as

$$\mathbf{F}_{tl} = \mathbf{F}_C + \mathbf{F}_{lat}, \tag{4.9a}$$

$$\mathbf{F}_{tl} = \frac{Wv^2}{gR} \mp W \sin\alpha \cos\beta. \tag{4.9b}$$

Referring to Fig 4.11(a), note that $\mathbf{F}_C$ is directed outward from the center of curvature, while $\mathbf{F}_{lat}$ is directed inward toward the center; thus, to make the sign of the tilt angle ϕ positive whenever the skier tilts toward the center of the turn, we adopt the sign convention used in Eq. (4.9b). When the skier turns past the fall line, or when the skier turns downslope at an angle β relative to horizontal that is greater than 90° [recall Fig. 4.5(a)], $\mathbf{F}_{lat}$ and $\mathbf{F}_C$ add together, as shown in Fig. 4.11(b). The expression for $\mathbf{F}_C$ given in Eq. (4.9b) inserts W/g for the value of the mass M. The total force $\mathbf{F}_{tl}$ and the gravitational force perpendicular to the snow plane $\mathbf{F}_N$ are orthogonal: they add as the square root of the sum of squares. This gives us

$$\mathbf{F}_{reac} = [(\mathbf{F}_C \mp W \sin\alpha \cos\beta)^2 + (W\cos\alpha)^2]^{1/2}, \tag{4.9c}$$

$$\tan \phi = \frac{v^2}{gR \cos \alpha} \mp \tan \alpha \cos \beta. \qquad (4.9d)$$

With the sign convention used, and for $v=0$, ϕ becomes negative, indicating that the ski rolls to the outside edge for the uphill quadrants and to the inside edge for the downhill quadrants of the turn.

Whenever the center of the carved turn is downhill from the skier's position, the skier must compensate for gravity and edge to the outside of the turn, on the uphill edge of the ski, while centrifugal force requires that the skier edge to the inside of the turn, on the downhill edge. If the skier's velocity is too low, the skier edges to the outside, turning uphill, or away, from the desired center of curvature. As the skier progresses around the turn, the edge angle increases, shortening the radius of the turn, perhaps even decreasing the radius so much that the total lateral force exceeds the bounds of the strength of the snow needed to hold the edge of the ski, and the skier skids out of the turn and falls. In this case, the requirements for dynamic stability and the constant radius of the overly tight turn, which is controlled by the ski edge angle, are not compatible. If the skier picks up speed during the turn, the edge angle needed for stability would increase even further.

Skiers can control the balance of the opposing forces that act in the carved turn in several ways. For example, the skier may use angulation, which we discussed earlier and is illustrated in Fig. 4.6, to set the ski edge at different angles as needed. Compensating through angulation in the uphill quadrants of the carved turn to make the edge angle Φ large enough for the desired turn radius is difficult; compensating through angulation in the downhill quadrants is easier. Skiers may also shift the effective load point for the application of $\mathbf{F}_{load}$, the force of the snow on the ski, either by bending forward or by sitting back on the tails of their skis. The radius of the turn carved by the edged ski depends on the configuration of the ski edge in contact with the snow. Unloading a part of the ski so that it does not carve the snow surface shortens the contact length, which may change the radius of the carved turn or induce some lateral skid to change the turn radius. Skiers may also use unweighting, rotation, and lateral projection (transferring weight from one leg to the other) to overcome many of the problems with edging cited above, and we will consider these techniques in the next chapter. Also, note that there is always some skidding, which is usually done to increase control, involved in turning on skis, even in the process of carving a turn. Let us look at some of these processes in greater detail.

Refer to Fig. 4.12, which shows the track of a parallel skier carving through a circular turn, a J turn, and a ''comma'' turn made in succession. The inside track gives the motion of the skier's CM; the outside track with the ski shapes superimposed upon it illustrates the path taken by the skier's feet along with the orientation of the ski as it moves through the turn. The direction of the ski tip indicates a load transfer from one ski to the other.

At point A, the skier initiates a circular turn to the right by running the ski to the left and tilting the body so that, when the center of mass starts into the turn, the ski is on the outside to balance the centrifugal force. At this point, the turn is only partially carved. The ski is not directed precisely in the direction of motion, and the skid increases as the turn progresses. This

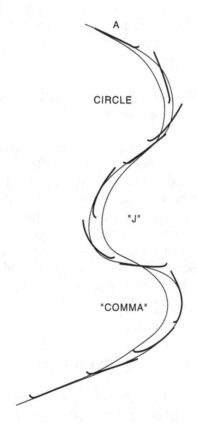

FIGURE 4.12. The ski orientations needed to carve the configurations for circular, J, and "comma" turns are shown by the track with ski shapes superimposed upon it at different points on the path. The direction of the ski tip on the track indicates the edge of the ski that contacts the snow. The angle of the ski to the track indicates the degree of skid. The inside track that the ski does not follow represents the path taken by the skier's CM.

skidding is necessary because the ski-print trajectory must cross the CM line ahead of the next turn in order to rotate the body tilt from right to left before the next turn starts.

For the J turn, the radius decreases as the turn progresses, so the ski print indicates the necessary increase in the tilt angle. In this case, the skier lets the tail of the ski release at the point of minimum radius to generate excessive tail skid. At the next step in the J turn, the tip of the ski skids to the left to move the ski print to the left, anticipating the short turning radius that marks the beginning of the "comma" turn. The skier changes edges and, using proper angulation to set the edge of the ski effectively into the snow, forces the ski to carve through the comma turn with little or no skid. When we examine a ski print made in the snow during a carved turn, the ski tip is in a skid mode because it must first plough the track and set the guide groove that the rest of the ski follows and carves. Also, the ski tip flexes as it ploughs through the snow in the manner shown. Readers interested in greater technical detail regarding the physical processes of carving turns should refer to the discussions in Technotes 3, p. 203; 4, p. 205; 5, p. 208; and 10, p. 233.

In large measure, the physical capabilities of the ski itself dictate the kinds of maneuvers that a skier can perform. Earlier parallel-ski techniques were, in part, limited by the use of earlier skis, which were relatively stiff and had almost no sidecut. Carving turns with such skis—at least as we understand carving today—was just not possible. As modern skis have become shorter and have offered a wider selection of flexure characteristics with generally greater sidecuts (culminating in the radical sidecuts of the recently introduced parabolic skis), carving turns has become progressively more feasible as the preferred method of turning. The general theme of Stenmark's comments cited above at the beginning of this section is that by 1990, ski instruction methods had not yet caught up with the capabilities of modern ski equipment [4]. With continued changes in ski equipment, we may expect instructional methods to continue to change as well, if more slowly.

SNOWBOARDING AND THE CARVED TURN

With the development of snowboarding, the technique of carving turns through edging and angulation has been highly developed with spectacular results. The interplay of carving and skidding needed to control a turn at high speed can be more easily observed in snowboarding. Most turns carved on skis are at tracking angles of a value β between 60° and 120° or within 30° of the fall line, so the severe edging problems that the snowboarder in

Fig. 4.13 solves by using extreme angulation to initiate and link the tightly carved turns shown simply do not arise in most skiing. An intermediate skier can learn much about the angulation techniques used to carve controlled turns on skis by looking down from the lift at the maneuvers of a snowboarder carving turns on the slope below.

Snowboards are designed to optimize the boarder's ability to carve a turn. The snowboard's sidecut generates an edge radius that, with the large angulation angles possible, allows the boarder to cut turns with very small radii. Also, the snowboarder's two-footed stance across the board and the board's flexure properties make it an extremely versatile turn-carving instrument, well suited for carving turns in either hard-packed or soft snow. Referring to Table 3.5 from the preceding chapter, note that the edge radii of snowboards are at least four times smaller than the edge radii of the skis listed in Table 3.4. The edge angle Φ of a snowboarder carving a turn may go up to 70°, so the carve radii of a snowboard may be as little as 3 m. Note also that on the asymmetric board listed in Table 3.5, the heel-side or backward edge radius is smaller than the toe-side or front radius. A snowboarder using this asymmetric board can lie out forward almost flat and recover his balance; that same maneuver is not so easily performed on a backward layout,

FIGURE 4.13. An expert snowboard instructor carves a tight turn. Note the angulation of the snowboard starting with the heelside at the top of the turn and finishing with the toeside at the bottom. (Reconstructed from video images of P. Naschak, PSIA Demo Team, used with permission.)

however. Look again at the snowboarder illustrated in Fig. 4.13 carving a turn in a sequence from a backward lean through the transition to a forward lean. The angulation in the forward lean is much larger than the angulation in the backward maneuver.

FREE-HEEL OR TELEMARK SKIING

All of the wedging and parallel-ski maneuvers that are performed with conventional alpine ski equipment can be performed equally well with free-heel skiing equipment. The basic telemark maneuver consists of a forward shuffle of one ski in a manner similar to the shuffle used in diagonal track skiing. In the forward shuffle, the skier unweights the ski so that it may be rotated to point across the tip of the weighted ski at some angle θ. At the same time, the skier's body comes forward to weight the forward ski. On a groomed slope, the forward shuffle should be relatively short. If the two skis are equally weighted, the stance resembles an asymmetric wedge with the forward ski edged somewhat more than the aft ski, which suggests that the telemark turn is essentially a type of wedge turn. If, on the other hand, the forward ski is fully weighted and angulated, the unweighted trailing ski plays almost no role in the turn, and the turn becomes a carved turn controlled entirely by the inside edge of the forward ski. If the skier's weight is not fully over the forward, carving ski, the skier will allow the angulation to relax during the turn so the tail of the weighted ski releases and skids out to some degree. This action, however, makes it difficult to recover rapidly when the trailing ski needs to be shuffled forward to set up the next turn. If the skier's weight is distributed to some degree on both skis, to carve a proper turn the skier's body must be angulated in the reverse-shoulder, counterrotation position, so both skis are edged effectively. When more than one ski carries the load, there will always be some combination of carving and skidding through a turn.

We can estimate the radius for a wedged telemark turn by looking at the geometry of this maneuver (see Fig. 4.14). Assume that the lead ski in the illustration is half its contact length ahead of the trailing ski, and it is edged into the snow so that the lead ski tracks at the angle θ to the trailing ski. If the full load remains on the trailing ski, the path is unchanged. If the full load sifts to the lead ski, the path of the skier suddenly changes into a new direction that tracks at the angle θ. If the load is balanced equally on the two skis, a curved track with the radius R will result. The path will follow approximately the arc of the circle that goes through the center of the afterbody of the trailing ski and the center of the forebody of the lead ski. The offset Δ then determines the radius of the circular arc of the turn by the

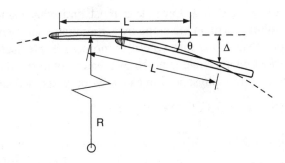

FIGURE 4.14. Geometry for skis turning in the telemark or free-heel position (top view). The trailer ski tip is placed at the center of the lead ski and at the angle θ.

relation $R = L^2/2\Delta$. Thus, for skis with a contact length L of 2 m and a value for θ of 15°, R, the turn radius becomes about 5.2 ms.

One common fault of free-heel skiers when they attempt to execute telemark turns on a groomed ski slope is that they let their trailing ski separate from their leading ski; in such a configuration, it is difficult to maintain stability. A forward shuffle that produces a lead of more than half the contact length of the ski is not advisable. One way to avoid this problem is to let the knee of the trailing leg rest against the calf of the leading leg. This procedure has the added advantage of edging the leading ski through enhanced body angulation, which enhances the carving action. The skier's upper body must be kept upright to keep the skier's center of mass over the midpoint of the effective contact area of the skis. And, finally, note that the lead ski can equally well be turned by the angle θ away from the direction of the trailing ski, and a similar turn will result. The Polish-American skier, Vic Bein, calls this maneuver the "Polish Telemark" [5]. However, this turn is not at all a new turn. Veteran skiers who can recall the days of Christiania turns will recognize Bein's "Polish Telemark" as the maneuver once designated the "Open Christy."

CONCLUSION

In this chapter we have discussed the basic mechanics of the most widely used downhill ski maneuvers. A competent skier must, however, link and adapt these maneuvers to suit the demands of the terrain and the conditions encountered to make the completed ski run a satisfying experience. Much of the skill needed to do this with confidence and success involves biomechanical feedback—"feel"—which can be learned only by practice. In the final analysis, ironically, no amount of analysis can produce an unerring formula for satisfying skiing. Each skier must find what maneuvers he or she can or cannot perform on the slope by skiing that slope again and again. Fortunately for most of us, that repetitive process is precisely the fun of skiing. In the next chapter, we analyze in greater detail the dynamic interaction of some of the physical forces at work when we attempt some common skiing maneuvers.

REFERENCES

1. For more information on Newtonian mechanics, readers may find useful the discussions in almost any college-level textbook that offers an introduction to physics. For example, see D. C. Giancoli, *Physics*, 4th ed. (Prentice-Hall, Englewood Cliffs, NJ, 1995), D. Halliday, R. Resnick, and J. Walker, *Fundamentals of Physics*, 4th ed. (Wiley, New York, 1993), or P. Hewitt, *Conceptual Physics*, 2nd. ed. (Addison Wesley, Cambridge, MA, 1992).
2. For a similar discussion of this problem, see R. Perla and B. Glenne, "Skiing," in *Handbook of Snow*, edited by D. M. Gray and D. H. Male (Pergamon, Toronto, 1981).
3. J. Howe, *Skiing Mechanics* (Poudre, LaPorte, CO, 1983).
4. I. Stenmark, "Ski Technique in the 1990s," in Snow Country, **1990** (March), p. 18.
5. V. Bein, *Mountain Skiing* (The Mountaineers, Seattle, WA, 1982).

CHAPTER 5

INTERACTIVE DYNAMICS OF ALPINE MANEUVERS

In the preceding chapter, we analyzed idealized situations in which the edge of the ski would bite into a plane or inclined snow slope that otherwise did not deform. In each case we assumed that the skier always anticipated lateral forces and maintained balance throughout the maneuver being discussed so that the total reaction force of the skier's center of mass was always directly along the skier's legs, through the skis, and onto the snow. In actual skiing, skiers alter the configuration of their bodies and, by using muscular forces, set their centers of mass in motion in relation to their skis. A simple example of this is the way skiers weight and unweight their skis by thrusting their legs up or down. More extreme examples would be setting the poles (or even a ski) into the snow and then moving or rotating all or part of the body over or around that stationary point in acrobatic jumps that may or may not be accompanied by twists and turns. In this chapter we examine several of these dynamic skiing maneuvers [1].

SKIING NONPLANAR SURFACES

The natural development of a planar ski slope that receives extensive skier traffic is the growth of moguls, the "bumps" or mounds of snow that form as skiers make their way down the slope, linking turn after turn. Only through regular grooming can the surface of a ski slope remain planar. Even mogul-designated slopes must be periodically groomed; otherwise, bare ground would soon be exposed in the troughs between the moguls. Mogul skiing necessarily involves a combination of skid turns with forced rotation and unweighting. Speed control, which, as we have seen, may be achieved in part through skidding the skis, is also necessary in mogul skiing. It is impossible to link pure, carved turns when the edge angle of the ski Φ changes rapidly because of the changes in the slope caused by the presence of moguls.

When skiing moguls, the skier should unweight—that is, release the edge of the ski—at the sides of the trough where the normal force is reduced and skid the ski using counterrotation of the body, which will set the opposite ski edge into the snow and initiate the turn. The combination of the absolute slope inclination and the reduced concave curvature of the slope at the top of a mogul, along with the minimal centrifugal loading that occurs, defines the physical conditions that make the top of a mogul the preferred location for turning. At the bottom of a mogul, the centrifugal loading through the skis and onto the snow is at its maximum because of the high curvature of the surface. Simply stated, this physical analysis shows that skiing a banked turn off the top of a mogul is the easiest and most effective strategy for skiing a mogul slope, even if this approach does involve some skidding. Finally, consider that for the expert, mogul skiing can become a gymnastic exercise. Expert mogul skiers use jumping, angulating rotation and counterrotation, and lateral projection to change the direction of their skis rapidly. Exceptionally gymnastic skiers may hop from the top of one mogul to the next, pivoting upon landing to take off for the next mogul. Let us look more closely at some of the physical forces involved in this dynamic type of skiing.

Figure 5.1 shows a skier moving down the fall line and the forces that act on the skier represented as a physical system with its center of mass indicated by the circle located approximately at the skier's waist. Assume that there are no drag or snow friction forces present. F_I is the acceleration

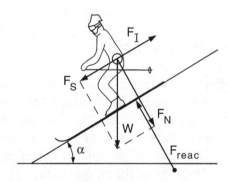

FIGURE 5.1. Skier on the fall line showing forces normal to the snow surface.

inertial force, that is, a force equal but opposite to $\mathbf{F}_S$, the gravitational force parallel to the slope. The location of the snow-reaction force, $\mathbf{F}_{reac}$, must coincide with the force normal to the slope, $\mathbf{F}_N$, to reduce the torque on the skier system to zero. Although a skier may use pole and ski placements to exert forces parallel to the surface of the snow, the skier in Fig. 5.1 does not do this; so, for this example, we consider only the normal reaction forces at the surface.

The gravitational force in the diagram is $\mathbf{F}_N$, which may be expressed as $W \cos \alpha$, where W refers to the weight of the skier and α represents the angle of the slope. Assume that the gravitational force, $\mathbf{F}_N = W \cos \alpha$, is constant. The reaction force of the snow on the skis, $\mathbf{F}_{reac}$, may be more or less than the gravitational force, depending on whether or not the skier thrusts his legs downward or retracts them. The difference between $\mathbf{F}_{reac}$ and $\mathbf{F}_N$ equals the inertial force from acceleration that is normal to the slope plane. Note that the center of mass of the entire system—the skier and all of the skier's equipment—is being accelerated and that only the external snow reaction force can cause that to happen. Thus the equation for the up and down motion of the skier's center of mass when it moves normal to the slope is

$$\frac{Wa}{g} = F_{reac} - F_N. \tag{5.1}$$

Recall from the previous chapter that the skier's mass M may be expressed as his weight W divided by the acceleration of gravity, or W/g; so Eq. (5.1) states that the skier's motion in the direction normal to the slope, Wa/g, equals the snow reaction force minus the normal force.

Now consider the example of a skier who moves his center of mass up and down as he skis over undulations in the slope, such as, for example, moguls. In this case, there may be both horizontal and vertical inertial forces that act relative to the slope and hence both horizontal and vertical accelerations that act relative to the slope. Figure 5.2 represents the motion of a skier's center of mass, shown as solid circles at the waist of the stick-figure skier, for this case.

When the curvature of the path of the skis is downward, as it is when the skier goes over the crest of a mogul, $\mathbf{F}_{reac}$ is less than $\mathbf{F}_N$, making the skier's vertical acceleration into the slope, $\mathbf{F}_{reac} - \mathbf{F}_N$, negative; so the skis unweight. When the curvature of the path of the skis is upward, as it is when the skier is in one of the troughs between a mogul, $\mathbf{F}_{reac}$ is greater than $\mathbf{F}_N$, so the skis press into the surface of the snow with greater force. These cases correspond to the points at which the skis are unweighting and weighting, respectively. The truth of these statements is immediately

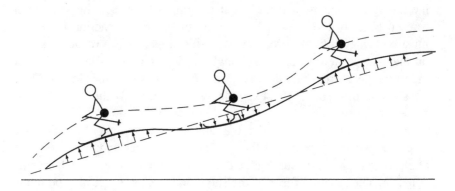

FIGURE 5.2. A skier in a fixed configuration on the fall line of an undulating slope. The distribution of the inertial forces normal to the slope is shown by the arrows.

evident on the hill when we ski up and down moguls in the manner shown in Fig. 5.2. Skiers can readily feel the increasing and decreasing of the pressure on their legs in this manner when they ski up and down moguls.

These statements can be put in a consistent form by recalling that the left-hand side of Eq. (5.1) (the mass represented by $W\mathbf{a}/g$) is the inertial force—the property of a mass body that resists the acceleration—that is always opposite in direction to the acceleration force $\mathbf{a}$. We can rewrite Eq. (5.1) with $\mathbf{F}_I = -W\mathbf{a}/g$ as

$$-F_I = F_{\text{reac}} - F_N ,$$

thus   (5.2)

$$0 = F_I + F_{\text{reac}} - F_N .$$

The sum of all forces on a mass body, including the inertial forces, must always equal zero. For the motion illustrated in Fig. 5.2, $\mathbf{F}_I$ is positive when the skier moves over the crests of the moguls (which unweights the skis) and negative when the skier moves through the troughs between the moguls (which weights the skis).

Now look at Fig. 5.3, which shows the motion of a skier moving across a trough and crest. In the trough, the skier extends her body to generate a constant reaction force on the snow, so that her center of mass (CM) does not accelerate downward; at the crest, the skier retracts her legs to maintain the constant reaction. Thus, in effect, the skier extends her legs going into the dip and retracts her legs on coming out of the dip to keep her center of mass moving in a straight line. When this is so, $\mathbf{F}_{\text{reac}}$ remains constant.

Up to this point in our discussion we have assumed that the skier's legs and skis have no mass. Of course, the legs and skis do have mass. So,

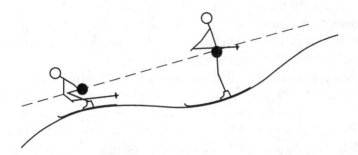

FIGURE 5.3. A skier on an undulating slope keeps the center of mass moving in a straight line; hence there are no inertial forces normal to the slope.

following the relationships between forces that we have just observed, when the skier's legs are tucked upward, the skier's body comes downward; conversely, when the skier's legs are extended, the skier's body goes upward. The relative motion of the skier's legs and body keeps the skier's center of mass moving in a straight line. This motion is all very natural to good skiers because they feel the forces we have described here as pressures on the soles of their feet, and they make the appropriate compensations, bending and straightening their knees, extending and retracting their legs, as they ski up and down, over and through the crests and troughs encountered on a slope with moguls.

ROTATION AND COUNTERROTATION

When we turn on skis, the skis must be freed from the snow either by skidding them or by, in some other way, unweighting the skis and releasing their edges. Then the skis must be rotated, and then, finally, the edges of the skis must be set into the snow in the new direction, locked into the new path by edge pressure and carving action. At the same time, the muscles of the body must absorb and counteract the lateral centrifugal force generated by the sudden change in direction. As we saw in the preceding chapter, the torque required to rotate the skier through a wedge turn is provided by the reaction forces of the snow on each ski. Let us look more closely at the rotational forces involved in turning skis.

The skier is an articulated body in which parts of the body may rotate and move relative to other parts of the body. For example, the shoulders and torso may rotate relative to the hips and legs. To rotate the skis through a turn, the skier applies an external, rotational force—a torque. Torque may be generated, for example, by planting a pole and rotating the body around it after releasing the skis from the snow by unweighting the edges, which leaves the skis flat on the snow and able to rotate. When the upper part of the body rotates relative to the lower part of the body, the unweighted skis must also rotate over the top of the snow.

To this point in our discussion we have considered in detail forces related to linear motion only. But, as we have just noted, skiers rotate their bodies as they ski. To describe those rotational movements, we need to refer to Newton's equations of motion for rotation. When a skier rotates his body, his CM is the origin of reference for the rotation. Various parts of the skier's body may rotate in various directions at various speeds and may have different motions, so each part that rotates will have its own angular momentum $\mathbf{L}$ relative to the origin of reference for the rotation, the skier's center of mass. We define angular momentum as $\mathbf{L} = r_\perp \mathbf{P}$. In the case for the skier seen as a system of component parts in rotation, $r_\perp$ is the perpendicular lever arm that defines a line from the origin of reference, the skier's center of mass, to the line that defines the direction of $\mathbf{P}$, the linear momentum of the component part. The total angular momentum for the skier system is the sum of the angular momenta—the vector sum—of all of the component parts of the system.

When we considered linear motion earlier, we referred to forces; for rotational motion, the analog of forces is torques. The equational expression for the torque τ corresponding to any force F acting at a point on the part in rotation defined by $r_\perp$ is $\boldsymbol{\tau} = r_\perp \mathbf{F}$. The direction of the torque is given by the right-hand rule: when $r_\perp$ rotates in the direction of $\mathbf{F}$, the torque that results has the direction of a right-handed screw rotating clockwise. Often in angular momentum problems the torque that acts is not in the direction of the angular momentum, so unexpected, seemingly counterintuitive phenomena occur.

The analog of mass, as we have been expressing it in our equations of linear motion, for rotational motion is the moment of inertia I, which we define as $I = \Sigma r_\perp^2 M$, where $r_\perp$ is the perpendicular distance from a reference axis through the center of mass to the mass element M and the values of $r_\perp^2 M$ are summed over all of the components of the system in question. The moment of inertia I will usually change with the direction of the axis of rotation. The angular momentum of a rigid body is $\mathbf{L} = I\boldsymbol{\omega}$, where is $\boldsymbol{\omega}$ is the angular velocity. For the sake of simplicity, assume that the skier consists of a system with only two parts and that each part is a rigid body. The skis, boots, and legs represent the lower part, which has angular momentum $I_1\omega_1$; the torso, arms, and poles are the upper part of the body, which has angular momentum $I_2\omega_2$. If the total angular momentum of this simplified skier system initially is zero, then

$$I_1\omega_1 + I_2\omega_2 = 0. \tag{5.3}$$

If the skier's skis are free to rotate, rotating the upper part of the body—the torso, arms, and poles—in one direction must cause the lower part of the body—the skis, boots, and legs—to rotate in the opposite direction.

Thus to initiate a turn to the right, the upper body should rotate to the left, that is, use counterrotation, which will produce a torque that turns the lower body and the skis to the right. Counterrotation works in this manner only if the skis are unweighted and sitting flat on the snow with their edges unset, completely free to rotate. This same principle explains why, when a freestyle skier launches into the air and performs a helicopter aerial, he rotates his upper body and arms in one direction to cause his lower body and skis to rotate in the opposite direction.

Using counterrotation to rotate the skis has its limitations. If there is substantial resistance to the skis' rotation, the skier must use forced body rotation techniques, such as the pole plant or ski plant. The force applied to an extended arm by planting the pole in the snow generates a torque that rotates the whole body. The skier couples the pole plant with unweighting of the skis to reduce the skis' resistance to the rotation. The ski plant with rotation (as in the wedge turn) applies torque through the hip joint to rotate the body and the ski.

In the parallel turn with skidding, the skier sets her body into angular rotation using pole or ski action to transfer angular momentum to the skis, causing the skis to skid sideways while they rotate. Such a turn is not very easily controlled; skiers are at the mercy of the slope and their own dynamics. As the skier comes around in the turn with the shoulders in the direction of the turn, the skier's weight will tend to move back and load the tails of the skis. Because the skis are already in a sideways skid, the skid is exacerbated by the loading on the tails, and the skier has difficulty restoring the weight loading over the center of the skis needed to retain control. We may readily observe this maneuver and its effects on the slope. When skiers who have not yet fully mastered the techniques of counterrotation and carving attempt to compensate for the excessive rotation of the torso caused by this maneuver, they will flail their arms and poles about, trying to recover their balance and stop the rotation.

LATERAL PROJECTION

Lateral projection is most useful when the radius of a turn, if it were completed by carving alone, would be too large to achieve the desired path. Lateral projection allows a skier to make tightly linked, small-radius turns.

For example, ski racers may use lateral projection to set the projected ski immediately on its inside edge to initiate each successively linked turn in a slalom. Using lateral projection helps racers gain added speed when they push off of their trailing ski going through a turn. Lateral projection is also useful at much slower speeds on the flatter portions of the slope or on a catwalk where skiers must skate their skis to maintain speed.

A skier may use lateral projection to set the unweighted, inside ski in the new, desired direction during a carved-turn maneuver. The skier exerts additional thrust by pushing off of the weighted, outside ski to accelerate the body mass in the new direction. In effect, the skier applies a sudden, additional lateral force that has a component along the new direction, so the skier's speed increases. One ski must be locked by its edge into a carving trajectory so the skier can use muscular torque to rotate her body and achieve the appropriate stance for each segment of the linked sequence of turns made using lateral projection.

LEG THRUST

Any body motion normal to the slope resulting from leg thrust provides some weighting or unweighting of the skis. For example, consider the simple up and down motion of the body that results from bending and straightening the knees to retract and extend the legs that we discussed to some extent above when we considered skiing moguls. The possible vertical displacement of a 6-ft-tall skier's center of mass through up-and-down leg thrust is about 14 in. For the up-and-down leg-thrust maneuver, and nearly all other leg-thrust maneuvers, the up-to-unweight and down-to-weight principles at work are illustrated in Fig. 5.4 using simple assumptions.

Readers interested in greater technical detail regarding the mechanics of unweighting and weighting should refer to Technote 6, p. 216, "Up and Down Unweighting." In Fig. 5.4, the action depicted occurs on a horizontal, planar surface over which a skier executes a turn that involves using up-and-down leg thrusts to weight and unweight his skis. The skier's center of mass is depicted by the circle at the stick-figure skier's waist. The skier's velocity is 20 mph or about 30 ft/s. The reaction time is 0.2 s. We assume that the up-thrust and down-thrust forces are constant. The down thrust can be only as large as the weight W when the skis do not touch the snow, so the skier's acceleration is $-g$. The up thrust is assumed to be $2W$ as a maximum, so the net upward force may be as large as W with an upward acceleration of g.

In Fig. 5.4(a), the skier starts the unweight–weight procedure in an up

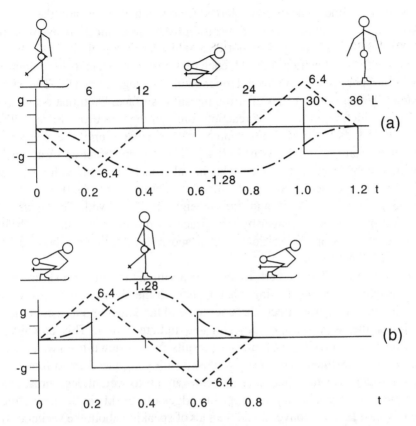

FIGURE 5.4. Weighting and unweighting skis using leg thrust and knee action. (a) shows a weight–unweight sequence; (b) shows an unweight–weight sequence.

position with the legs extended, from which the skier can retract his legs. The skier ends the unweight–weight sequence in a crouch, and the recovery weight–unweight sequence returns the skier to the leg-extended stance. At the start of the sequence shown, the skier retracts his legs for 0.2 s and accelerates downward at the rate g, the acceleration of gravity. The skier's velocity becomes -6.4 ft/s, and the displacement of the skier's center of mass is -0.64 ft. At this point the skier must thrust his legs downward to stop his fall. We assume that the net up-thrust force is also W for 0.2 s, after which the skier's body is at rest, but the displacement of the center of mass is now -1.28 ft.

The unweighting occurred immediately during the first 0.2 s, at which point the skier set the inside edges for the turn. At the end of the turn and before the next turn, the reverse process must take place so the skier can prepare for the next maneuver. A larger leg retraction would permit a longer unweighted time. These conclusions follow whether or not the forces involved are constant. The time integral of the forces must always be zero because the skier's body does not change its vertical motion.

Figure 5.4(b) illustrates the sequence for an up-to-unweight case in which the skier begins the run in a crouch with his legs retracted. The skier extends his legs and pushes off from the snow surface. Up thrust occurs for 0.2 s at which point the snow reaction force on the skis drops to zero. The skier then falls freely with no change in leg extension until the skis touch the snow again, which happens in 0.4 s. At this point, a net up thrust stops the motion and returns the skier to his initial crouch position, with his knees bent and legs retracted. Notice that the unweighting occurs for 0.4 s and that the skier rises to 1.28 ft, and the unweighting is delayed. The unweight interval may be made longer by leg retraction; however, the final up thrust must occur to stop the vertical motion and return the skier's body to its original configuration.

For the most effective performance in carving a turn, modern ski instructors usually recommend using what appears to the skier to be a downward motion: the instructor directs the skier to bend her knees and retract her legs to lower the skier's center of mass and initiate a turn. This downward motion of the skier's center of mass results from a down-to-unweight leg thrust that immediately unweights the skis, allowing the skier to rotate the skis through the turn and then execute an up-to-weight leg thrust that completes the turn, setting the opposite edges of the skis in the new direction. Notice that the conventional manner of speaking about the sequence of maneuvers involved in this turn, "down–turn–up," describes what skiers physically experience as their center of mass lowers and then raises, but it is not correct physics for describing the forces at work. When a skier sinks into a crouch at the beginning of a turn, she does not push herself downward, she pulls her legs up. With the legs and skis retracted, the skier's body inertia decreases the snow reaction force and unweights the skis, freeing the edges from the snow momentarily until they fall back down to the snow surface. When the skis return to the snow surface, the skier's body mass has a velocity component downward that the legs absorb by further knee action (a downward leg thrust, which results in an upward thrust off the surface of the snow) until the skier's vertical velocity becomes zero and another turn is ready to be initiated.

Skiers may also use basic, up-and-down leg-thrust maneuvers to minimize the time that a skier will be out of contact with the snow surface—that is, airborne—as a result of skiing at high velocity over uneven snow surfaces. If a skier's center of mass moves in a straight line without regard for the configuration of the slope, the force on the skis will be constant and the skier may use techniques appropriate for skiing a plane slope. To keep the center of mass moving in a plane, the skier must contract and extend the legs—use basic, up-and-down leg-thrust maneuvers—so that no centrifugal forces act upon the skier normal to the slope. As speeds increase, however, there is a limiting downward curvature for which it is impossible to maintain contact with the snow. Centrifugal force depends on the velocity squared. Without proper anticipation, unweighting the skis to initiate a turn on an uneven snow surface at high speed can be disastrous: the skier loses contact with the snow, making recovery very difficult. Simply put, if you ski faster and faster over uneven terrain, you will eventually unweight to initiate a turn, become airborne, and likely fall rather than reset your edges and recover your equilibrium. The rule of thumb is "do not lose contact with the snow surface," so let us examine pretucking in anticipation of a break in the slope, which is one of the options a skier has for minimizing the flight time that the skier will be airborne.

When a skier's centripetal acceleration going over the curve at the brow of a slope break is larger than the acceleration of gravity g, becoming airborne is inevitable. In Fig. 5.5, we see a simplified example of a slope break in which we assume that the initial slope is horizontal and that the break in the slope makes an angle δ of 15°. If a skier runs off this break in the slope at a speed of 44 ft/s without anticipating the break in any way, the skier's landing point is the horizontal distance $2d_m$ to the right of the takeoff point C at the brow of the slope break, which is expressed as

$$2d_m = \frac{2v^2}{g} \tan \delta. \tag{5.4}$$

If, however, the skier anticipates the break in the slope and prejumps it, launching herself into the air with a downward leg thrust in a manner high enough and early enough so that she lands at the break C, tangent to the slope to the left, the skier's trajectory is the path shown by the left portion of the dashed curve to the left of the break point C. The airborne distance

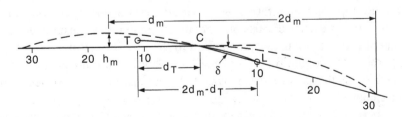

FIGURE 5.5. Prejumping to shorten airborne distance at a slope break.

is again the value $2d_m$, given in Eq. (5.3). If the skier cannot launch herself high enough into the air to rise to the height h_m, she will bounce before C and become airborne again. Thus the dashed curves give the limiting trajectories.

If the skier were to raise her center of mass to the height of h_m and then tuck to unweight at the distance d_m to the left of the break point C, the skier would land at C and follow the slope to the right of C. Launching horizontally at the point d_T before the slope break C and from a height h achieved by tucking gives a total flight distance of d_h, which we may express as

$$d_h = 2d_m - d_T. \tag{5.5}$$

If d_T is zero, the net result of this maneuver is the same as running the slope break. If, however, $d_T = d_m$, then $d_h = d_m$. To achieve this minimizing result, the skier must tuck from a height of h_m. For the situation illustrated above—a slope break with an angle of 15° that the skier runs at 30 mph or 44 ft/s—the optimum tuck height for minimizing the flight distance becomes 2.2 ft, and d_m, the flight distance itself, becomes 16.2 ft. Two feet is about the maximum a person could tuck. Running off of the break at full speed with no anticipatory maneuvers results in a flight distance of 32.4 ft with a minimum flight time of 0.37 s. The skier needs about half a second just to decide to tuck, so if the skier wishes to minimize flight distance when taking off from a slope break, she must anticipate the slope break by about 40 ft and then execute either an up or down leg thrust—tuck or prejump—in the appropriate place prior to reaching the break in the slope. Readers interested in a more detailed analysis of these techniques should see Technote 7, p. 219, "Analysis of Prejumping."

BODY ANGULATION

Body angulation requires rotation or counterrotation of the upper body relative to the lower body: the hips, legs, and skis. Counterrotation of the upper body, which results in the upper body being turned so that it faces down the slope, is the preferred technique for creating effective body angulation. In our previous discussion of counterrotation, we considered it as a means of providing angular momentum to the skis. In the present context, the purpose for counterrotation of the upper body is not to impart angular momentum to the skis, but to enhance body-angulation which increases the skier's control.

The photograph presented earlier in Fig. 4.6(b), p. 92, illustrates the static configuration of proper body angulation that results from counterrotation. Refer to Fig. 5.6, which illustrates a skier using counterrotation to create effective body angulation as he moves through a full turning cycle. Notice how, through counterrotation of the upper body, the skier uses body angulation to project his hips toward the slope, which increases the angle on the inside or upslope edges of his skis while he maintains his center of mass over the center of his skis or on the inside, as required to compensate for centrifugal force. Using counterrotation, the skier turns his head and upper body throughout the turn to look down the fall line, allowing him to anticipate his next maneuver.

As we mentioned in the preceding chapter, counterrotation is sometimes called "reverse shoulder." From the movement depicted below, we can see why. The skier's outside or lead shoulder is brought backward to a position from which the skier could look back over that shoulder up the slope. On a turn to the left, the right shoulder moves backward, opening the upper body so that the skier faces down the fall line as the skis run across the slope to the left. On the ensuing turn to the right, the counterrotation of the skier's upper body and resulting angulation of the skier's upper and lower body reverses, and the left shoulder moves backward.

Less proficient skiers tend to bend forward at the waist, rather than keep their upper bodies vertical to the slope and use counterrotation and body

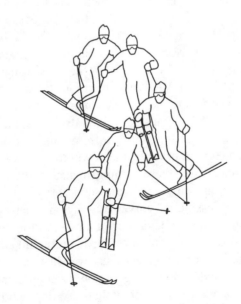

FIGURE 5.6. Sequence of body positions during a full turn cycle illustrates counterrotation or the reverse-shoulder maneuver.

angulation to control their balance through the weighting and unweighting of their skis in a series of linked turns. Such a forward lean with the knees flexed relatively little may cause the tail edges of the skis to release into a skid that must be overcome before the skier can initiate a new turn. So do not lean forward. Let counterrotation and body angulation swing you through your turns.

INDEPENDENT LEG ACTION

Independent leg action is an integral part of both the older and the more modern techniques for turning skis. Skiers learning to turn are commonly advised to "ski on the inside edge of the outside ski." This advice suggests that skiers should ski, in effect, on one ski (the inside edge of the outside ski), leaving the unweighted inside ski to be used as it may be needed, perhaps to set a new turn direction (through lateral projection) or to compensate for lateral instability (the skier might weight the inside ski to regain lost balance). Before we look at how independent leg action maneuvers figure in modern techniques used to carve turns, we first consider how independent leg action works in another, less advanced technique for turning skis.

When beginning skiers learn to negotiate the slope using the wedge or snowplow technique, they manipulate their skis somewhat independently, but both skis are always weighted (that is, their edges are set into the snow surface) to provide braking or torque action. When a beginner starts to make the transition from wedge turns to parallel turns and parallel skiing, the skier stems the uphill ski (lifts it from the snow using independent leg action) and sets it down in the new direction, transferring weight while skidding the ski to complete the turn in a maneuver that combines carving and skidding (see Fig. 5.7). The skier might also thrust the downhill ski out to brake or to serve as a stable platform from which to slide the uphill ski into the new turn direction, bringing the downhill ski to carve parallel to the outside ski. In either case, these turning maneuvers use independent leg action, and the resulting turn is called a stem christie.

In our earlier discussion of carved turning in Chap. 4, we noted the instability zone present at the beginning of a carved turn: the need for lateral stability requires that skiers weight the outside or uphill edges of their skis, but in order to initiate a turn, skiers must weight their inside or downhill edges. The use of independent leg action through lateral projection— dynamic transfer of weight with or without skating action—can help the skier overcome the uncertainties encountered in the instability zone.

To use independent leg action effectively, the skier must be completely

FIGURE 5.7. Independent leg action during a wedge turn. The right ski wedges to initiate the turn, and then the weight transfers to the left or downhill ski at completion.

comfortable skiing with only one ski weighted, that is, skiing on the inside edge, or the more difficult outside edge, of one ski only. Doing this requires perfect lateral canting: the ski must run flat on the snow when the skier's center of mass is over that ski. The skier carving perfect turns down on a smooth slope with her skis locked tightly together may look to be using both skis working together as a single unit. In reality, that skier must ski each ski independently—she must transfer her weight from one ski to another through independent leg action—to be able to handle the bumps, crud snow, soft snow, and the whole range of other variable terrain conditions that she may encounter, even on a groomed slope.

CANTING

To use independent leg action effectively, the skier's knee joint should be more or less directly over the centerline of the ski when the ski lies flat on a horizontal surface. This placement ensures that leg loading of the ski acts perpendicular to the ski's running surface. If a skier is bowlegged, the running surface of the ski may be tipped laterally when the skier stands naturally. To ride the skis flat on the snow, the bowlegged skier can compensate for this problem by separating his legs to a relatively greater degree or by flexing his knees and rotating them to change the lateral angle of his skis. In either case, transmitting the dynamic reaction loading from the center of mass to the inside edge of the outside (carving) ski may be difficult because of a lateral torque on the knee and ankle. The preferred solution to this problem is canting: modifying the ski, binding, and boot alignment so that the ski rides flat on the snow without the skier's having to make any of the biomechanical adjustments to his knees and legs that we described above.

Several commentators have noticed the role that canting plays in skiers' ability to edge their skis and carve turns [2]. Canting may be especially important for women because their wider pelvic configuration and thus larger hip socket separation suggests that women, even without any trace of bowleggedness, must ride their skis with a greater separation between them than would a man of the same stature. If this suggestion is at all correct, cants could, potentially, help any woman carve turns more easily.

It is fairly easy to check one's skis to see if canting would be beneficial. The skier stands in his boots and bindings and skis on a flat surface with his knees flexed. A plumb bob extending down from the skier's knee should fall about a 0.25 in. to 1 cm inside the centerline of the ski. A quick check can also be performed in the field. Standing on a flat section of the snow-pack, the skier or a companion places a straightedge across the top of the skis. If the straightedge does not lie flat across the skis, the degree of misalignment and the required adjustment will be quickly apparent. Significant misalignment may be corrected by adjusting the pads or cuffs in the boots, by placing insoles in the boots, or even by grinding the bottom of the boot to offset the desired lateral angle. Sometimes lateral wedges, called cants, may be applied to the bindings to achieve the desired degree of lateral offset that will place the ski exactly flat on the ground.

Skiers with an inclination to adjust their own equipment to suit the special demands of their anatomy will find canting one of the easier and quite possibly most effective adjustments they can make to their gear. The importance of canting has been recognized among a few skiers for many years. More recently the general skiing public has become aware of canting as more and more ski shops will now, as a regularly offered service, evaluate skiers for canting and perform needed adjustments.

FORE AND AFT SHUFFLE

The advice to "ski the inside edge of the outside ski" would seem to suggest that the relative fore and aft positions of the skis should matter little. Not so. On a diagonal traverse, the uphill, inside ski must lead the downhill, outside ski. The edge angle of the downhill ski will be the greater of the two, so it will tend to carve uphill and into the track of the uphill ski. To counteract this tendency, the skier must skid the downhill ski slightly by shuffling it fore and aft across the surface of the snow as needed to maintain the track of the turn and the skier's balance.

When beginning a turn, it is natural for the weighted, outside, downhill ski to be a little ahead of the unweighted, inside, uphill ski. As a skier's weight shifts to the back of his skis toward the end of one turn, the unweighted, inside ski moves forward. At the point where the next turn is about to be initiated, the skier has the option of using the unweighted, inside ski to steer the turn. Using lateral projection, the skier may transfer his weight to the new outside ski with a "shuffle and step" action whereby the inside ski shuffles ahead and into the path of the turn.

ARM AND POLE ACTION

The pole plant enhances the up–down–up body action recommended by some older ski instruction methods for unweighting the skis and initiating a carved turn. The more modern, down–turn–up technique for carving turns that we described above does not require planting the poles, so today one seldom sees expert skiers really planting their poles into the snow as they ski down well-groomed slopes. When skiing off-piste or in the backcountry, however, the pole plant still has a place to enhance the complete release of the skis from soft or crud snow. Also, the pole plant helps when skiing over terrain where jump turns are essential, such as when skiing moguls, steep and icy patches, or in narrow gullies.

Poles serve other, less dramatic, functions as well. Skiers are seldom seen on the slope without poles; most of us would feel lost skiing without our poles. Poles offer the means for applying additional traction force—we can push off of them—at the beginning of a run or when we come out of a skid turn in which we have lost speed and we need to impart increased momentum. In these instances, the poling force that can be applied may be substantially larger than the available gravitational traction, especially on a shallow traverse that avoids the fall line. The poles may help us restore our balance in an emergency, and they may be used almost like ice picks to give us some adhesion when our skis are not biting into the snow well enough to hold their edges.

But more than the poles themselves, the general attitude of the skier's hands and arms when holding the poles affects the skier's balance and speed control. The hands and arms represent a considerable wind drag area, so they should be held in front of the body if the skier wants to increase speed. For this reason, the poles used for speed skiing are curved so that their baskets trail in the lee of the body when the hands are held in front. The ski racer usually has the trunk of his body bent well forward in a tuck with the arms stretched forward. When recreational skiers try to emulate the racer's stance and they bend forward at the waist, they usually do not have the technique, strength, and quickness necessary to ski effectively in that position. The recreational skier should position his upper body so that his center of mass is directly over his boots and his arms are outspread. Just as tightrope walkers extend their arms and hands out on either side for balance, so using a similar position for the arms and hands (and poles) helps the skier maintain lateral balance. The moment of inertia of the upper body about the axis of rotation vertical to the skis through the boots will be much larger in this configuration. When initiating a turn, the torque applied to the skis and boots by counterrotation of the upper body will be maximized, and the outstretched arms allow ready pole touch for balance as needed. Finally, the increased wind drag factor offers skiers a means of speed control when they carve turns that use little or no skid. The aerodynamic drag that results from skiing with the arms outstretched in this manner may be easier to handle than skidding for speed control.

Even beginning skiers must synchronize many techniques into a coordinated sequence of events to negotiate even the most modest slopes successfully. The synchronization of techniques, such as the weighting and unweighting of the skis so that they carve and skid accordingly, the rotation and counterrotation action that results in body angulation, and the fore and aft shuffling of the skis through independent leg action, resists extensive quantitative analysis. The synchronization process requires internal, qualitative, biomechanical feedback. Once again, in the final analysis, skiers must learn to feel the difference between successful and unsuccessful performance by paying careful attention to what happens when they are out skiing on the slopes.

REFERENCES

1. For qualitative descriptions of the dynamic processes at work when we ski, see J. Howe, *Skiing Mechanics* (Poudre, LaPorte, CO, 1983); G. Joubert, *Skiing an Art... a Technique*, (Poudre, LaPorte, CO, 1980); G. Twardokens, *Universal Ski Technique* (Surprisingly Well, Reno, NV, 1992).
2. For more information on canting, see C. Carbone, *Women Ski* (World Leisure, Boston, MA, 1994), pp. 106–111, who discusses at length the special appropriateness of canting for women skiers; see also W. Witherell and D. Evrard, *The Athletic Skier* (Athletic Skier, Salt Lake City, UT, 1993), pp. 19–65.

CHAPTER 6

HIGH-PERFORMANCE SKIING

Many aspects of skiing have been analyzed in the context of competitive, high-performance skiing. The term *high-performance skiing* commonly refers to ski racing, speed skiing, acrobatic skiing, ski jumping, and extreme skiing, usually in association with organized, high-level competition. In this chapter we will look at a few issues related to high-performance, competitive skiing that should, nevertheless, prove interesting to recreational skiers. For example, any factor that decreases velocity, as aerodynamic drag does, is crucially important for the high-performance racer or speed skier. Although aerodynamic drag is usually not an especially important issue for recreational skiers, it affects the recreational skier in the same manner that it affects a racer or speed skier. For that reason, all skiers may achieve some greater understanding of their performance—whether it be high, medium, or low—by understanding something about the way aerodynamic drag affects a skier, which is the first high-performance skiing issue we will consider.

AERODYNAMIC DRAG

For the racer or speed skier, the effect of aerodynamic drag is probably larger than the effect of snow-friction drag, which makes it an extremely important factor. Readers interested in a more detailed discussion of the technical aspects of aerodynamic drag should refer to Technote 8, p. 222, "Aerodynamic Drag" after finishing the discussion below. Aerodynamic drag always acts on the skier at some effective point determined by the size

127

and shape of the body area the skier presents to the wind stream. For example, even when a ski jumper positions his body in an extreme forward lean with his exceptionally wide skis spread apart to enhance his aerodynamic lift, drag forces are still induced at various points on his body. The ski jumper's extreme forward lean accentuates lift, but it also helps him compensate for the pitch induced by drag, lift, and gravitational forces. Let us look more closely at the problems presented to downhill racers by aerodynamic drag.

There are two types of aerodynamic drag. The first is form drag, which results from a difference in air pressure across the skier's body. Form drag is far and away the most significant aerodynamic drag factor for a skier. The second type of aerodynamic drag, boundary layer or skin drag, contributes only about 2% of the total aerodynamic drag on the skier. The characteristic, tucked position of the downhill racer represents an attempt to assume a truly aerodynamic shape, that is, a shape with a long body dimension parallel to the wind stream. Researchers working for the U.S. Olympic Committee and the Canadian National Ski Team have conducted wind-tunnel tests designed to analyze the effects on aerodynamic drag produced by variations on the traditional, tucked position [1].

The speed of a downhill skier is easily over 60 mph, or 88 ft/s. The record velocity for speed skiing is about 145 mph, or 213 ft/s. These speeds are well in excess of the minimal speed needed to create a turbulent boundary layer between the skier's body and the wind stream, which reduces the drag factor caused by differences in air pressure. The Canadian researchers found that a low, egg-shaped crouch in which the body is folded against the thighs and the arms are pressed against the chest with the lower legs almost vertical produces very low aerodynamic drag. The skier tries to hold his thighs parallel to the slope with his back rounded between 20° and 60° to the slope. If the skier extends his arms down toward his boots, the drag nearly doubles. If the arms must be lowered from the body, they should be aligned in front of the legs, not beside them. If a skier could hold his arms straight out in front of his body, the effect would be to streamline his overall body shape and lower drag would result. That position, however, is almost impossible to hold for any appreciable length of time in a race.

Wind-tunnel tests have illuminated other drag factors related to ski equipment. For example, the upturned tips of skis contribute about 7% of the total aerodynamic drag on a skier, and that is why perforated ski tips that allow the wind stream to pass through them are now available. Because downhill races are generally on hard-pack snow, the profile of the ski tip can be significantly reduced. Similarly, rear-entry boots with smooth front surfaces help reduce drag. The size of the baskets on the ski poles used in racing are also being reduced, even though they do not seem to have much effect on aerodynamic drag.

The surfaces of helmets and ski suits have long been assumed to be optimal when they are smooth, and the composition, texture, and porosity of the materials used in manufacturing ski suits for racing have been prescribed by international agreements. There is reason to believe, however, that a puckered surface for helmets and suits may reduce aerodynamic drag, but this hypothesis has never been tested. Some bobsledders, for example, put sandpaper over the front of their helmets in an attempt to reduce the aerodynamic drag on their helmets.

No matter how advanced our understanding of the most optimal tucked position may become, aerodynamic drag will always be a major factor in downhill racing because the tucked position is possible only on selected segments of any downhill course. Downhill racing demands weighting, edging, and body angulation to stay on course, and none of those maneuvers can be done effectively in a fully tucked position. Streamlining ski clothing and equipment may, in the long run, be more effective responses to our developing understanding of skiing and aerodynamic drag.

QUICKEST DESCENT TRAJECTORIES

Much of the strategy in a downhill or giant slalom race involves choosing the optimum route down the slope. This is the classic brachistochrone problem (from the Greek *brachistos,* meaning shortest, plus *chronos,* meaning time): discovering the path of shortest time of descent down an inclined plane. In addition to choosing the quickest route, the giant slalom racer must also select optimal turn radii and judge where best to initiate those turns relative to the slalom gates [2].

Both the downhill and the giant slalom races are exercises in skiing in the shortest possible time through a series of relatively long segments between control points (poles or gates) which require a change in direction. During the race, most of the elapsed time is spent running the segments, not turning through the gates. Thus the racer's first objective is to achieve the highest possible speed skiing between gates. Skiing as directly down the fall line as possible achieves the highest speed, but such a course will not get the skier across the slope to the next gate. Thus the skier must optimize the motion down the fall line for speed while at the same time he must select the shortest distance across the slope to the next gate.

The answer to this problem shows that taking a longer, curved path—called the cycloid—and skiing initially close to the fall line in order to generate higher speed results in a shorter transit time than is required to traverse directly across the slope, in spite of the greater distance traveled between the two gates in the slalom. In this instance, a straight line between two points, while it is the shortest distance, is not necessarily the shortest time between those points. For a full, mathematical development of this problem and its solution, see Technote 9, p. 229, "The Brachistocrone Problem: The Path of Quickest Descent."

TURNING WITH MAXIMUM SPEED

During most of a race, a skier's speed is limited primarily by snow friction and aerodynamic drag. During turns, however, carving (to a lesser degree) and skidding (to a much greater degree) increase the snow-friction forces on the skis and diminish the skier's velocity. The problem facing the high-performance skier is to turn with minimal deceleration. It is true that speed can be increased slightly in a turn by employing a pumping motion with the body as the skier comes around the turn. This pumping maneuver is discussed below and in greater detail in Technote 10, p. 233, "Pumping To Increase Velocity." The more effective strategy for increasing velocity, however, is to decrease snow friction by carving through the turn as much as possible on the edges of the skis rather than skidding. Winning time margins in major races are of the order of a couple hundredths of a second measured over a span of 60 or more seconds, so very small differences in technique may determine the winning margin. The hundredths of a second that might be saved or lost by carving rather than skidding in a turn are crucial to a racer's success.

One tactic for making fast turns is called "going straight–turning short" (GSTS) [3]. The emphasis in this case must be on *carving* the turn; any skidding during the turn represents a loss of kinetic energy to the work done by the skid. Refer to Fig. 6.1, which shows the trajectory down a slope of 25° for a slalom course with three turns (at points B, C, and D) run by an idealized skier who enters the course at point A with a low initial velocity of 15 km/h. The idealized skier represented in this model makes instantaneous turns through the gates to create an optimal Z trajectory, which assumes that no friction acts to slow the skier. The idealized turns represented in the figure are assumed to be as short as possible, and in the limit they represent the skier's use of some combination of ski and pole plants that involve no energy loss and allow him to make discontinuous changes in the velocity vector at the gates marking the turns in the slalom course. Some pumping action through the turns may also increase the skier's speed. Coming out of turn B, the skier's speed is 40.29 km/h, at point C his speed is 54.97 km/h, at point D he reaches 66.48 km/h, and he finishes the slalom

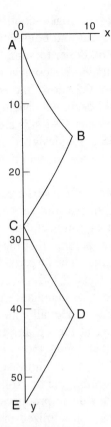

FIGURE 6.1. The layout of an optimal Z trajectory with the horizontal and vertical distances given in meters. The transits AB, BC, CD, and DE are identical, each with the interval $x=7.5$ m and $y=13$ m on a slope of 25°. [Adapted from G. Reinisch, "A Physical Theory of Alpine Ski Racing," Spektrum Sportwissenschaft I, 27 (1991).]

at point E traveling 76.28 km/h. The increases in the skier's speed that result in this model are excessive; they reflect the idealized, frictionless state we assume for this case. Notice, however, how the segments of the skier's run between the poles become straighter as the skier's velocity increases from one segment to the next.

The straight trajectories of the idealized GSTS turns illustrated in Fig. 6.1 are impossible to achieve in reality because, among other things, a ski's sidecut inevitably causes the ski to carve into the slope and thus away from the fall line—in effect, the sidecut makes the ski run uphill. Skis are stabilized by riding the inside edge of the outside (downhill) ski, and most skis have some sidecut to enhance their carving action. Running a straight line requires running the ski flat against the snow, but a ski running in this manner would lose all lateral—or edge—control. Designers of racing skis must balance the need for sidecut and edge control with the need for flat running and speed. This can be done, to some extent, if the ski's edge is elliptical so that the section under the boot is straight but the tip and tail have a large edge taper. Figure 6.2 shows the edge profiles of some recreational skis and of a racing ski designed specifically to make GSTS turns. The Dynastar Omeglass and Volant FX1 skis are examples of conventional ski designs, the K2 Biaxial GS ski is designed specifically for giant slalom racing, and the K2-4 ski has a radical sidecut that gives the ski an hourglass shape.

Note that the tail contact points of the skis are aligned at 0 cm and that the centerline for the contact length and the shovel contact points are given. Note also that the shovel contact points and the positions of greatest width are not exactly the same, and remember that the boot loading position on a ski is always behind the centerline. The larger sidecut slopes of the fore and afterbody portions of the K2 Biaxial GS ski combine with the relatively straight edge section under the boot to enhance the ski's ability to perform the GSTS maneuver. Running with a small edge angle between the ski and the snow, there is little loading at the tip and tail, so the ski runs on the straight edge under the boot, allowing the skier to track in the desired direction without skidding while maintaining stability and control. At the large edge angles needed for turning, the tip-to-tail loading is large because the ski is flexed, and the radical sidecut produces a short turn radius, which generates a tight turn. Skidding is minimized in both the straight runs and

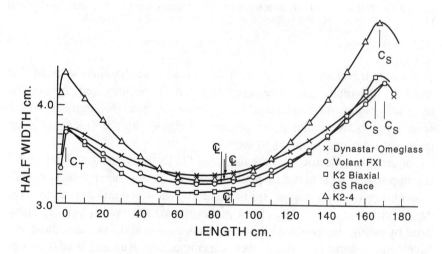

FIGURE 6.2. Edge profiles of several skis.

the turns, and the skier is better able to "go straight–turn short" through the course.

The sidecuts of the different skis illustrated in Fig. 6.2 tell us something about the evolution of ski design and its effect on the evolution of ski technique. The Volant FX1 ski is a fairly recent, high-performance recreational ski. The Dynastar Omeglass is an older model, recreational ski. These skis have sidecut radii of 44.8 and 58.3 m, respectively. Given these sidecuts, there is simply no way that any skier could carve very short radius turns on either of these skis without some skidding, although the Volant, with its shorter sidecut radius, would carve a noticeably shorter turn than the older Dynastar model. In contrast, the K2-4 ski is an example of the very-narrow-waisted, parabolic sidecut skis for recreational skiers that have only very recently appeared on the market. The K2-4 ski has a sidecut radius of 25.6 m, which means it can carve turns with very short radii, in much the same manner that a snowboard carves a turn. In fact, some skiers using such skis with radical sidecuts can, like snowboarders, ski without poles, laying out over the snow and touching it with their gloves as snowboarders do when they angulate their bodies through extremely tight, very-short-radius turns. For both competitive racers and recreational skiers, design and technique evolve in a constant interplay of innovation and response.

PUMPING TO INCREASE VELOCITY

Consider how a snowboarder riding at a very low velocity in the bottom of a half-pipe formation with nearly vertical sides can pump his body up and down and, in a couple of pumping cycles, ride up to the crest of the sidewall. Some of the snowboarder's increased velocity is achieved by setting the board's edge and pushing off against the snow, a technique similar to what we have called lateral projection. But in addition to the acceleration produced by lateral projection, the snowboarder's pumping action may occur in a direction perpendicular to the snow's surface, which adds to his kinetic energy and thus to his velocity. Skiers can use a similar pumping action to increase their velocity when they ski in a trough or through undulations in the slope.

To understand how pumping with the body can increase velocity, consider how a child on a swing pumps herself higher and faster. Note that the child in motion on the swing is isolated from her surroundings except for the radial forces that apply at the point from which the swing is suspended. There is no work done by the external suspension system itself—whatever the swing is attached to—because it does not move.

Because no outside force acts to do work on the system, we may well wonder how it is that the state of motion changes as the child pumps herself up to a higher swing velocity. The answer to this query is that the child's pumping motion does work that adds kinetic energy to the system.

The principle of the conservation of energy tells us that to increase the velocity of the swing system, work must be done. Work is the product of a force and a displacement in the direction of the force. For the case of the child in the swing, the force is the centrifugal force, which is an inertial force directed radially outward. There are several ways to pump the swing. One common method used when standing on a swing is to be in a crouch at the bottom of the arc of the swing and then, as the swing moves, stand up rapidly, crouching again when the swing comes to rest at the highest point in the cycle, and then standing up rapidly again when the swing has its maximum velocity, that is, as it passes through the lowest point in its cycle, creating an up-and-down pumping action that repeats as the swing goes through its complete cycle. By standing up, the child moves her center of mass in such a way that her body works against the centrifugal force, increasing the kinetic energy of the system. On the following excursion, the swing comes to rest at a higher elevation, which represents greater potential energy for the next cycle. On the next cycle, the velocity increases because the swing falls from the greater height. In this case, work has been done on the system internally, not externally, and that work increases the energy in the system and the velocity of the swing. For a more detailed discussion of how energy generated by pumping with the body may be converted into kinetic energy, see Technote 10, p. 233, "Pumping to Increase Velocity."

In a similar manner, a skier may increase his velocity by having his center of mass as low as possible going into a trough and then rising rapidly at the bottom of the trough when the centrifugal force is the highest and the pumping action will do the greatest work. Continuing over the top of a hump, the skier may likewise crouch down, lowering his center of mass to allow the gravitational force to work against the centrifugal force and increase his velocity again, so long as he does not crouch so fast that his skis unweight and lose contact with the snow. During the crouching motion, gravity helps the skier pull his body downward, and gravity has the opposite effect during the skier's rising motion; so gravity does no net work during one of the complete pumping cycles.

Pumping in the plane of the slope may also increase velocity, although the fractional increase in kinetic energy will not be as large as it is when the skier goes through a trough. Pumping action and the forward and backward pivoting motions that it may be combined with are illustrated in Fig. 6.3. Pumping action, coupled with a forward-and-backward pivoting of the body

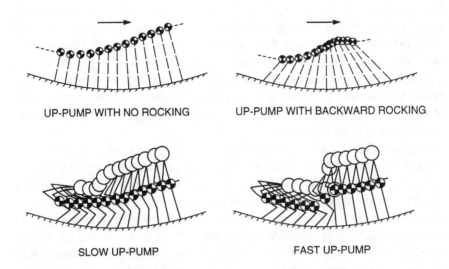

UP-PUMP WITH NO ROCKING   UP-PUMP WITH BACKWARD ROCKING

SLOW UP-PUMP   FAST UP-PUMP

FIGURE 6.3. Up-pumping used in four combinations to increase velocity (Mote and Louie, 1982).

in which the skier's leg thrust is synchronized with the instant when the centrifugal force is at its maximum, enhances the work done and increases velocity.

BALANCING AND EDGING ON ONE SKI

Throughout our discussions of skiing dynamics, we have analyzed skiers in what may be called the steady state of motion; in most cases we have not investigated how dynamical transitions occur from one moment to the next. For example, in our discussion of the unweighting and rotation maneuvers needed to initiate a carved turn, we assumed that the skier's body was leaning at the proper angle toward the center of rotation so that all of the resultant forces passed through the contact line at the edge of the ski. In reality, the skier's body assumes that position (or fails to assume it) in response to dynamic feedback generated by the biomechanical systems of the body's musculature and nervous system as they react to the forces generated as the skier performs the turn. Analyzing how this process works is a classic problem in modeling dynamic control systems. Many activities have been analyzed in this manner, including skiing [4].

As we have seen, modern ski turning is done primarily on the inside edge of the outside ski. This means that a skier, balancing on one leg and one edge during a turn, may be modeled and analyzed as an inverted pendulum.

To achieve useful results with this model, however, we must neglect the fact that the skier's body is articulated at the waist and assume, for the purposes of our analysis, that the skier's body is rigid. The inverted pendulum model that results allows us to suggest how a skier's biomechanical feedback systems interact with the competing physical forces that come to bear on the skier during a turn. Readers interested in a fuller development of this discussion should see Technote 11, p. 236, "The Skier as an Inverted Pendulum."

In our earlier discussions of turning, the skis always appeared in steady state of motion models placed at just the correct position they should be at throughout the progress of the turns under discussion. With a dynamic model, we see more clearly how a skier must react to the forces generated during a turn and actively move his body and his skis to maintain equilibrium. Figure 6.4 plots the lateral displacement of a skier's center of mass and skis as the skier moves through a turn. The example is for a skier who is 2 m tall and whose center of mass is at the mid-point of his body, 1 m up from the ski. The distance traveled in the initial direction of motion is given by L; D (the solid line) is the lateral displacement of the skier's center of mass projected onto the inclined plane; and D_1 (the dashed line) represents the motion of the ski. The distance between the two plotted curves, D and D_1, is the projection of the angle at which the skier must bank his body to achieve equilibrium as he moves through the turn.

The skier initiates the turn at the zero point given in Fig. 6.4. For the first

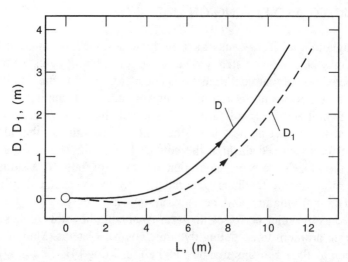

FIGURE 6.4. Lateral displacement from the initial direction of motion of the center of mass and skis as a skier moves through a turn. [Reprinted with permission from J. M. Morawski, "Control Systems Approach to a Ski-Turn Analysis," J. Biomech. **6**, 627 (1973) (Elsevier Science Ltd., Oxford, England).]

2 m of the turn, his ski turns in a direction opposite to the turn, giving an offset of about 0.25 m before the skier's center of mass starts into the turn. Note that this offset increases somewhat as the turn progresses. The separation between the ski and the skier's center of mass represents a tilt angle of about 30° during the turn illustrated. At a velocity of 10 m/s, the skier will need 0.4 s to initiate the turn illustrated in the figure. As the skier's speed increases, the distance traveled during the initiation of the turn may increase up to a total of about 10 m.

Initiating a turn from a straight track on a single ski means skidding or carving to force the ski outward, or unweighting the ski with hip angulation so that it can be rotated and then set in the new direction. At the point where the skier's center of mass starts to move into the arc of the turn, the ski must be edged to carve the arc and provide the needed centrifugal reaction force. This maneuver can be accomplished by projecting the inside ski laterally and on edge to the appropriate displacement for the next turn. In order to ski with one's feet close together, getting that outside ski out at the correct separation involves a pole plant, unweighting, and rotation; these maneuvers must be done over a reaction time of about 0.2–0.4 s. Biomechanical measurements of athletes show the lower leg reaction times of 0.2 s; for the upper body, reaction times of 1 s are common. This suggests that a skier skiing at a high speed of 20 m/s may travel as much as 4 m before initiating the combined upper- and lower-body maneuvers required to execute a turn. Although this discussion may not contribute directly to the improvement of one's skiing, these reaction times do point to some of the physical limitations that affect a skier's ability to anticipate and react to the forces generated during a turn executed at high speed.

EDGING, SKIDDING, AND CARVING

In our earlier discussions of the carved turn, we assumed for the sake of simplicity that the ski carved uniformly throughout the turn. Field observations of even the most accomplished skiers indicate that this cannot be true. Even the most purely carved turn must involve some skidding. Several studies have used idealized models to analyze in detail the motion of a skier moving from initiation to completion of a carved turn [5]. These models assume that the turn has a constant radius and is executed at a constant speed on a horizontal surface, which allows prediction of the orientation, flexure, and edge loading of the skis throughout the turn. An imposed force is applied to keep the speed constant. The ski itself is modeled as a beam that has a distributed bending stiffness in flexure, but is rigid in torsional and lateral stiffness. The edge profile of the ski, represented by its sidecut radius, taken with the bending of the ski produced by body angulation, determines the edge loading on the snow. The deformation strength of the snow's surface is modeled using measurements generated by cutting ice surfaces at different rake angles with a cutter similar to the edge of a ski.

For a known force normal to the plane of the snow surface (F_N in earlier discussions), the model uses the maximum value of the force that the snow may sustain in the plane of its surface perpendicular to the ski edge to calculate when the ski is skidding. These data determine whether a particular segment of the ski edge carves or skids at a particular point in the turn. In addition, the model gives values for the lateral forces that are generated by the snow.

A simplified version of the geometry of a skier making a turn is shown in Fig. 6.5. The skier is modeled as a rigid body with the plane of the ski always fixed perpendicular to the line that extends from the skier's center of mass. Thus the ski rotates laterally with the skier's center of mass, but there is no forward or backward rocking motion, neither is there any up or down pumping motion. In the model, any friction forces generated by motion parallel to the ski edge are neglected. Because the surface of the snow is horizontal, only the skier's gravitational weight acts to generate the normal forces that load the ski edge and press it against the snow. The efficiency of the turn is measured by the amount of the imposed force needed to keep the velocity of the center of mass constant throughout the turn.

Because the forces acting at the ski edge may change during a turn, they produce a lateral torque, which we assume the skier compensates for by

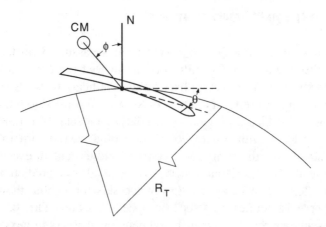

FIGURE 6.5. Geometric model of the skier in a turn with the radius R_T.

changing his tilt angle (ϕ) and by applying a forward to backward torque through his boot. The ski tracks around the turn to the outside of the trajectory of the skier's center of mass, so the ski turns, or yaws, toward the inside at an angle θ relative to the direction of motion, as shown in Fig. 6.5. As the skier moves through the turn, the ski flexes in varying amounts along its length, causing different segments of the ski edge to yaw at varying angles relative to the direction of motion. Given these conditions, the failure strength of the snow determines which parts of the ski edge will skid, that is, have motions perpendicular to the local orientation of the edge, pushing the snow before it; and which parts of the ski edge will carve, that is, move tangentially to the edge, making a longitudinal groove in the snow.

An idealized ski that carves perfectly through a turn would have no yaw angle; all parts of the ski would move in the same curve or arc, exactly parallel, directly in the carved track. The model we have just described more nearly approximates what happens in the field: there is no ski whose edge can completely carve through a turn. In all actual carved turns, some parts of the ski's edge will yaw under loading as the surface of the snow deforms and fails beneath it, and those parts of the edge will skid rather than carve. Remember that the local force at the edge of the ski normal to the snow surface varies widely along the contact length of the ski. This very detailed model leads to complex, nonlinear equations that must be solved numerically to generate conclusions. In the discussion that follows, we consider qualitatively some of the observations about the dynamics of a carved turn that may be drawn from this model.

Figure 6.6 shows a ski as it progresses through three stages in a carved turn. Assume that our model skier initiates a turn, either by unweighting or by lateral projection, and his load bearing, outside ski is in the skid stage. Figure 6.6 shows the geometric aspect of the ski on the snow plane. Because of the skier's body angulation and forward weight transfer in Fig. 6.6(a), loading appears at the inside edge of the ski's tip, and the forebody of the ski flexes and twists as it skids laterally and also bites into the snow surface, which causes it to develop even more flexure and forebody loading. The skier starts moving in the arc of the turn, and the trailing parts of the ski begin to move into the groove created by the ski tip and forebody, doing so without as much of the lateral skidding motion experienced by the tip and forebody. Remember that an imposed force equal to the component of the skid forces must be provided along the direction of motion to keep the skier's velocity constant; otherwise the speed of the skier would decrease. In Fig. 6.6(b), the afterbody of the ski begins carving the turn, tracking with minimal skid in the groove created by the ski tip and forebody, even as the forebody of the ski continues to skid. Finally, in Fig.

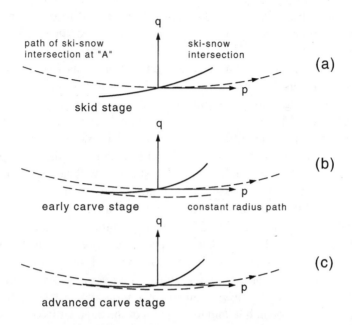

FIGURE 6.6. Skidding and carving of a ski during a turn as the turn progresses from (a) the skid stage to (b) the early carving stage, to (c) the advanced carving stage. (Lieu and Mote, 1985. Copyright ASTM. Reprinted with permission.)

6.6(c) only the tip portion of the ski continues to skid, generating the groove in which the remainder of the ski carves.

Now let us consider how the distribution of force along the ski edge changes as the ski moves through the three stages of the carved turn described above. In the Fig. 6.7, the y axis represents the force on the ski and the x axis represents the length of the ski edge with the forebody and the afterbody of the ski extending to the right and left of the center point, designated by the 0.00 value. In Fig. 6.7(a) the ski is in the full skid stage as the turn is initiated. The forebody edge is heavily loaded by the skier's weight-forward attitude. During the transition from skidding to carving shown in Fig. 6.7(b), the load distribution remains dominant on the fore-body edge, but it has begun to move back onto the afterbody edge. Finally, in the carving phase [Fig. 6.7(c)], the load is dominant at the center point and extends in a sharply smaller force over the complete afterbody edge all the way to the tail of the ski. In this final stage of the carved turn the forebody edge is still skidding. There is, however, relatively little or no loading on it because the ski flexes under the boot as it impinges into the snow to generate the carve groove, causing the forebody to be deflected

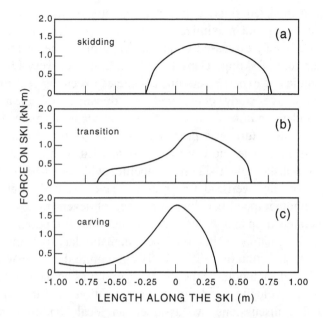

FIGURE 6.7. Force distribution along a ski edge. (Renshaw and Mote, 1991. Copyright ASTM. Reprinted with permission.) View (a) is the force distribution for a skidding ski; view (b) is for a ski in transition from skidding to carving, and view (c) represents a fully carving ski.

upward, so it may make little or no contact with the snow and thus generates little or no edge force.

The particular bending stiffness distribution used for the model ski in this example accounts for the large loading that appears under the boot in the graph above. Not all skis would exhibit this bending stiffness distribution and behave exactly in this manner. These results, however, usefully illustrate what happens when an actual, carved turn is executed on the slope.

FIELD STUDY OF CARVING TURNS

It is very difficult to determine from the analysis of models, no matter how detailed and complex the models may be, the change, if any, in the velocity of an actual skier that results from the skidding forces generated during a carved turn. In principle, the more a skier carves through a turn, the less energy is dissipated and, presumably, the faster the skier travels; conversely, the more a skier skids through a turn, the more energy is dissipated by the edge of the ski cutting across the surface of the snow, and the skier slows down. But the models do show us that some skidding takes place in any carved turn, so the change in velocity for an actual skier, as opposed to a modeled skier, will depend greatly on the skier's ability to minimize skidding and maximize carving. The problems associated with this effort are widely discussed in qualitative terms in the professional and popular literature of skiing. There have been, however, very few quantitative field studies of expert skiers on real slopes executing real turns.

One quantitative study considered the dynamics of an actual turn performed under conditions similar to those that occur in a giant slalom race [6]. A giant slalom turn was prepared on a 35° slope with a salted and compacted surface. The snow surface was marked with a grid. The skier was an accomplished giant slalom competitor with an expert's skill level. Observing from an overhead tower as the skier went through the turn, researchers photographed the skier at a speed of seven frames per second, which generated a spatial description of the skier's motion, that is, the geometry or kinematics of the turn. The kinematic data from this study are given in the first section of Table 6.1. By computing the traverse angle and the turn radius at each point along the skier's path as he moved through the turn, the skier's velocity and tangential acceleration were determined.

In preceding discussions, we assumed an ideal condition: that a ski sliding parallel to the ski's edge experienced no drag force. The only force occurred when the ski slid laterally or perpendicular to the ski's edge, which acts as a chisel. Now we calculate the component of the skid forces tangential to (F_{treac}) and lateral to (F_{lreac}) the path of motion of the skier's center of mass. The tangential forces associated with the ski's carving in its track were computed at each of the seven observation points. Forces are positive in the direction of motion and negative if they act against the direction of motion. The force parallel to the ski track, F_p, is the component of gravity along the skier's trajectory. Readers may recall that we discussed F_p previously in Chap. 4 [see especially Fig. 4.5(b), p. 90] and in Technote 3, p. 203. The drag force F_D consists mostly of snow friction on the ski and some aerodynamic drag. Because the skier's acceleration before the turn is zero, the value for the drag force F_D is therefore a negative value, equal to the component of the gravitational force along the track, and it is assumed to be constant throughout the turn. The observed tangential acceleration yields the inertial force, which allows the total tangential reaction force generated by the snow, F_{treac}, to be computed along the path of motion.

Lateral edge forces are positive if they are directed outward from the center of rotation and negative if they are directed inward toward the center of rotation. F_{lat} is the lateral component of the force of gravity. The observed radius of curvature gives the centrifugal inertial force F_C. The total lateral reaction force generated by the snow, F_{lreac}, is the negative sum of F_{lat} and F_C.

TABLE 6.1. *Giant slalom turn mechanics.*[a]

	Before turn	0/4	1/4	2/4	3/4	4/4	After turn
				Gate			
Kinematic data							
Velocity (ft/s)	46	43	35	35.5	36	35	35
Acceleration a (ft/s^2)	0	−12	−13	0.6	0.6	−0.7	1.8
Traverse angle β (deg)	22	30	61	90	116	147	152
Turn radius R_T (ft)		26	30	33	36	41	
Tangential edge forces							
$F_p = W \sin \alpha \sin \beta$ (gravity in lb)[b]	31	41	71	81	73	44	38
Drag force/F_D (constant)	−31	−31	−31	−31	−31	−31	−31
Inertial force $(-Wa/g)$[c]	0	53	57	3	3	3	−8
Total tangential reaction force F_{treac}	0	−63	−97	−47	−39	−6	−1
Lateral edge forces (skidding)							
$F_{lat} = W \sin \alpha \cos \beta$ (gravity in lb)	−76	−71	−39	0	36	68	72
Centrifugal force F_C $(= Wv^2/gR_T)$	0	314	180	168	159	132	0
Total lateral reaction force F_{lreac}	76	−243	−141	−168	−195	−200	−72
Total reaction force $F_T = (F_{treac}^2 + F_{lreac}^2)^{1/2}$	76	251	171	174	199	201	72
Yaw angle θ ($\tan \theta = F_{treac}/F_{lreac}$)	0	15	35	16	11	5	0
CM tilt angle ϕ ($\tan \phi = F_T/F_N$)[d]	−33	65	56	56	60	60	32

[a]Data taken from Glenne and Larsson, 1987.
[b]W is the skier's weight, $=142$ lb. α is the slope of the hill, $=35°$. β is the traverse angle from the horizontal.
[c]g is the acceleration of gravity, $=32.2$ ft/s^2.
[d]F_N is the weight component normal to the slope, $= W \cos \alpha = 116$ lb.

By adding the tangential and lateral edge reaction forces, we find the total edge reaction force F_T, which is the total snow-reaction force on the ski in the plane of the snow surface. Except for the point before the turn initiation, these total reaction forces are directed to the skier's left, toward the center of the turn. Assuming that the ski is oriented essentially perpendicular to the total edge reaction force throughout the turn, we can compute the yaw angle θ. Notice that the ski attains a yaw angle of as much as 35° (at point 1/4), but the angle drops rapidly to a value of only 5° at point 4/4, toward the end of the turn. The large yaw angle at the initiation of the turn accounts for the skier's deceleration as the ski skids into the turn and the skier slows from 43 ft/s at point 0/4 to 35 ft/s at point 1/4.

Refer now to Fig. 6.8, which offers a schematic illustration of the skier as

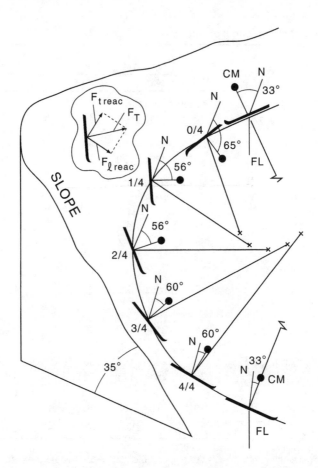

FIGURE 6.8. Geometric representation from field studies of a skier turning past a gate in a giant slalom. (Data from Glenne and Larsson, 1987.)

he moves through the seven stages observed in the turn. The snow exerts almost no force parallel to the edge of the ski. Thus the total edge reaction force must be perpendicular to the edge of the ski, which we note in Fig. 6.8 by the orientation of the ski at a yaw angle of 35° relative to the track at point 1/4. The snow reaction force normal to the slope plane plus the total edge reaction force sum to determine the snow reaction force on the skier. To maintain his balance, the skier must tilt his center of mass (CM) by the angle ϕ to align it with the snow-reaction force.

The line labeled N in Fig. 6.8 represents the normal to the slope plane. Before the turn, the skier tilts to the right from the direction of N to be in a vertical stance; thus he rides the uphill edges of his skis. Upon initiation of the turn, the skier rotates his body by 98° from right to left for stability.

The racer observed in this field study had to skid his skis over the snow to initiate the turn and get his skis to carve. Undoubtedly the forebody of the ski was highly flexed as it skidded into the turn, probably much more than the 35° shown for the average yaw angle over the first half of the turn. Over the latter half of the turn, note that the ski carved with little deceleration. Also over the latter part of the turn, the skier's center of mass probably moved backward to load the afterbody of the ski as it carved through the turn.

So the advice to "accelerate out of a turn" means, in reality, that once the skis begin to carve, the skier stops decelerating, which likely feels to the skier as if he is accelerating. The results of both the computer model of a carved turn and the field observation of a giant slalom skier carving a turn demonstrate that some skidding must occur in the early part of a carved turn. Common sense also tells us that this must be true when we see snow debris spray out from under a racer's skis as he skids over the snow to initiate a turn.

REFERENCES

1. For discussion of the tests sponsored by the U.S. Olympic Committee, see M. S. Holden, "The Aerodynamics of Skiing, Technology of Winning," in Sci. Am. **258**(2), T4 (1988); the Canadian National Ski Team studies are reported by A. E. Raine, "Aerodynamics of Skiing," Sci. J. **6**(3), 26 (1970).
2. See G. Reinisch, "A Physical Theory of Alpine Ski Racing," Spektrum Sportwissenschaft **1**, 27 (1991).
3. Reinisch, cited in Ref. 2, uses the "Going-Straight-Turning-Short" terminology and discusses the maneuver at some length in both qualitative and quantitative terms. See also G. Twardokens, *Universal Ski Techniques* (Surprisingly Well, Reno, Nevada, 1992), who also uses this terminology and offers an interesting qualitative discussion.
4. Much of the discussion that follows here and in Technote 11 comes from J. M. Morawski's analysis of the skier as an inverted pendulum in his article, "Control Systems Approach to a Ski-Turn Analysis," J. Biomech. **6**, 267 (1973).
5. See, for example, D. K. Lieu and C. D. Mote, Jr., "Mechanics of the Turning Snow Ski," in *Skiing Trauma and Safety: Fifth International Symposium*, ASTM STP **860**, 117 (1985); A. A. Renshaw and C. D. Mote, Jr., "A Model for the Turning Snow Ski," in *Skiing Trauma and Safety: Eighth International Symposium*, ASTM STP **1104**, 217 (1991).
6. See B. Glenne and O. Larsson, "Mechanics of a Giant Slalom Turn," in *The Professional Skier* (Professional Ski Instructors of America, Lakewood, CO, 1987), Winter Vol. 3, pp. 23–26.

CHAPTER 7

NORDIC TRACK, CROSS-COUNTRY, AND ADVENTURE SKIING

The term ''nordic'' may be applied in general to any skiing not done on the groomed slopes of an alpine ski area. Nordic track and cross-country skiing are usually done on prepared tracks or on designated routes that are used regularly. A nordic course with a relatively wide track is designed for skiers who use the skating technique; grooves set in the snow off to one side of the track are designed for skiers who use the diagonal stride technique. Although it is true that cross-country skiing routes may not be specifically prepared, for the purposes of our discussion we treat cross-country skiing (sometimes called touring) as nordic skiing done on routes used regularly enough that the snow has been packed, creating, in effect, a prepared surface. Nordic track skiers never have to break a new trail through the snow; cross-country skiers must do so only on relatively rare occasions. Finally, the elevation gains typically encountered in track and most cross-country skiing are usually modest, and any extended or relatively steep runs downhill are usually configured to provide run-outs, so speed control is seldom a problem.

TRACK AND CROSS-COUNTRY EQUIPMENT

Table 7.1 lists the dimensions of some representative classical, freestyle skate, and cross-country, or touring, skis. Track racing skis are, in general, longer than other nordic skis. They have little or no sidecut because carving

TABLE 7.1. *Geometric properties of classical, freestyle skating, and touring skis.*

Ski make Model	Classical Fischer C1200	Freestyle skating		Touring			
		Fischer Skate 195	Fischer Revolution	Asnes TorboxT60	Dynastar Montange	Karhu 205	Tua Escape20
Dimensions							
Chord length Xp1 cm	193.0	193.0	144.0	211.0	196.0	200.0	197.0
Contact length C cm	174.0	174.0	124.0	185.0	174.0	177.0	179.0
Shovel width S cm	4.00	4.00	4.80	6.70	5.90	5.10	6.30
Waist width W cm	4.40	4.40	4.50	6.00	5.00	5.10	5.00
Tail width T cm	4.20	4.22	4.70	6.20	5.20	5.10	6.00
Sidecut SC cm	−0.15	−0.15	0.13	0.23	0.28	0.00	0.58
Center height Hc cm				2.60	2.50	2.10	2.10
Projector area Ap sq cm	769	770	611	1239	971	959	1031
Comments				Metal edge		Waxless	Metal edge

turns is simply not a part of track skiing technique. Nordic track racing skis may even have a negative sidecut—a bulge—making their widest dimension at the waist, not the shovel, as is the case for the Fischer C1200 listed in Table 7.1. The narrower tip helps the ski stay in the track. These skis also have a double camber that forms a wax pocket on the bottom of the ski under the foot where a kicker wax is applied. All of the racing models of nordic skis, whether track or freestyle, are quite stiff to give them directional stability because they are used exclusively on prepared tracks and do not need the softer flex of a ski intended to run on unpacked snow. All of these skis also have a guide groove on their running surfaces that gives the skis longitudinal stability.

Freestyle skating skis are usually somewhat shorter than other nordic skis. Their thick midsections give them their flexural stiffness and raise the skier's boots higher off of the snow so that they will not drag during skating. The inside edges of freestyle skating skis are reinforced to provide the strength needed to withstand the stress put on those edges by the skier's skating action.

Finally, the average touring ski used by recreational cross-country skiers is about the same length as the nordic racing ski, but it may be somewhat wider to allow for flotation in the softer, unpacked snow that tourers may encounter in off-track skiing. Touring skis have a modest sidecut that aids the skis' turning performance. Like track racing skis, touring skis have either a double camber, to create the wax pocket needed for proper kicking technique, or they have a specially prepared running surface under the foot that gives the skis kicker traction. Recreational touring skis may vary in stiffness, but they are never as soft as the backcountry skis that are designed to be used in the unpacked and powder snow that adventure skiers seek out. The dynamic properties of all of these nordic skis are not of much concern for our discussion, and for that reason they are not listed.

The bindings used on nordic skis are an integral part of the boot and ski system. Only recreational touring skis use the three-pin, toepiece binding that accepts the common, recreational touring boot. Track skis used for racing are fitted with a boot–binding system that is designed as a unit. In many models, a tongue and groove between the boot and the foot plate secures the ski alignment when the foot loads the ski. In a few other models, a transverse hinge pin at the toe secures the toe piece and preserves the ski alignment. Most nordic skis, whether racing or recreational, do not use bindings that have release systems because the skier's heel is not bound to the ski by the binding, which allows the leg to torque to a great degree independently of the ski without placing dangerous stresses on the ankle, knee, or hip joints.

The variety of styles available in nordic boots is greater than the variety of ski styles. For classical track skiing, the equivalent of a low-cut running shoe will suffice. The grooved, prepared track used by classical track skiers guides the skis quite well on all but the steepest downhill sections of the course. Track skiers do not need boots with the stiff ankle support that would give them the firm edge control needed to control speed and guide the ski through turns. The soles of track skiing boots are very flexible to allow the skier an extended forward lean. These boots are very lightweight to increase their racing performance. For the freestyle skate ski racer, the boot must provide ankle support. Skating skiers push off on the inside edges of their skis, so there must be enough torsional rigidity in the boot to prevent the foot and ski from rolling since the edge is not under the center of the foot. Freestyle skate skiers perform much like ice skaters, and they need the same lateral support in their boots. Freestyle skate ski boots extend up and over the ankle with lateral stiffening, and they have a stiffer sole to return the ski to the foot quickly during the skating maneuver.

The recreational cross-country skier requires a boot that gives good ankle support and has a reasonably stiff sole to pressure and control the ski. The boot should fit the foot well enough such that when the boot is laced the foot has little or no play but is still comfortable. Earlier models of molded, one-piece composition touring boots, even expensive ones, were notorious for suffering structural failure. Some of these earlier boots, with their one-piece construction, featured a pin plate molded into the sole. In some cases, the sole would fracture and the toe part would separate from the boot, leaving no means to attach the boot to the ski. In such an emergency, the only way to make a field repair is to wrap duct tape around the ski and boot and limp home. Fortunately, newer models of one-piece, molded composition boots appear to have solved these problems. Molded boots are stiffer than stitched leather boots, which means that, in general, they give better ankle support. Leather boots, however, with their stitched seams on the welt and sole, may fail, but they are not likely to fail so catastrophically that the skier cannot ski home.

Typically, the poles used for nordic skiing are long enough to reach to the top of the skier's shoulder for diagonal stride skiing (whether racing or touring) and to the midface for freestyle skate skiing. Poles used for freestyle skate skiing must be stiffer than those used for diagonal skiing because the skier applies larger forces to the poles with moments at the grips, which cause the poles to buckle. If a cross-country skier expects to negotiate any kind of a slope while touring, the ideal pole length is substantially shorter. When traversing a slope on a cross-country track, having a shorter pole on the uphill side is a great advantage. Given the variety of circumstances cross-country skiers may find, adjustable poles are particularly desirable.

NORDIC TRACK TECHNIQUES

All nordic skiing, if the skier expects to pursue it with some degree of seriousness, requires considerable training to develop effective coordination and sufficient stamina. In nordic skiing more than in any other sport or exercise activity, the body produces a large mechanical work output for relatively longer, sustained periods. The nordic skier's performance depends greatly on how well the skier's training has increased the work capacity of the skier's relevant muscles and, possibly even more significantly, increased the aerobic capacity of the skier's heart–lung system, which must provide the oxygen the body needs to perform this strenuous work. Nordic skiing is completely aerobic exercise; no part of it involves anaerobic exercise, that is, exercise for which atmospheric oxygen is not needed. When we view the nordic skier as a physical system, that system actively involves the skier's skis, boots, poles, legs, arms, torso, and head, along with the skier's aerobic oxygen supply system. This system is different for each nordic skier. As the U.S. Olympic Committee has found, if we want to differentiate nordic skiers from each other to observe their techniques and performance, we need detailed temporal as well as spatial studies of film clips to see what is happening [1].

Every skier, whether alpine, nordic, or backcountry, uses the diagonal stride technique at some time for routine propulsion. The nordic track, diagonal stride sequence illustrated in Fig. 7.1 is typical. Notice that one leg and the opposite arm–pole combination are synchronized. In frame one, the skier plants the right ski with his full body weight to depress the concave, wax pocket of the ski under the boot to give the ski traction with the snow. The skier plants the left pole simultaneously, which provides arm thrust on the opposite side of the skier's body. These two propulsion forces—from the ski and from the pole—are applied to opposite sides of the body, so no torque is generated about a vertical axis. Notice how the skier's shoulders remain essentially transverse to the track. In frame two, both skis glide as the left foot and the right pole come forward to make the plant pictured in frame three. Then, another glide occurs in frame four, and finally in the last frame, 1.37 s later, we see an exact replica of the first frame: the skier is in position to repeat the sequence.

On occasion, track skiers using the diagonal stride technique will double-pole, pushing off from pole plants made with both arms synchronized to develop equal thrust on both sides of the body. Because all of the thrust produced by double-poling is developed in the arms and shoulders, this technique is effective only in short bursts, and it is usually performed only on level ground or on a downhill portion of the track.

Because of the stop-and-go motion of the legs and skis and because the

FIGURE 7.1. A Professional Ski Instructors of America (PSIA) certification team member demonstrates the diagonal stride technique. (Reconstructed from video images of J. Aalberg, PSIA Demo Team, used with permission.)

length of the pole thrust phase is restricted, the maximum speed possible using the diagonal stride technique is considerably less than the speed that may be obtained using the skate-stride technique. The stop-and-go motion of the legs and skis and the vertical motion of the skier's center of mass that is required to generate the thrust needed to force the wax pocket under the foot of the ski into contact with the snow represents a considerable amount of energy expense that does not lead directly to forward motion. Nordic track racers who use the freestyle skating technique uniformly beat diagonal stride skiers by as much as 18%, so nordic ski races have to be classified as either classical (diagonal stride only) or freestyle (skating and other techniques are allowed). The diagonal stride technique, however, is the only stride possible when a ski tourer carries a backpack, breaks a new track in the snow, or skies for extended periods of time.

Several ski-skating techniques are used to ski in prepared tracks. Figure 7.2 lists the terminology used to describe the actions performed by the poles and skis when executing ski-skating techniques, which may be classified as every stride (ES) or alternate stride (AS), depending on whether the skis and poles execute the same motions on both sides or on alternate sides.

The ES technique, which is illustrated in Fig. 7.3, is the simpler method

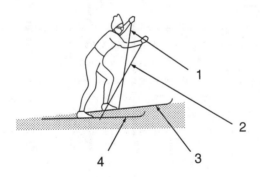

POLES:

| 1. | Glide Pole | Strong Side Pole | Hang Pole |
| 2. | Non-Glide Pole | Weak Side Pole | Push Pole |

SKIS:

| 3. | Poling Ski | Strong Side Ski | Glide Ski |
| 4. | Non-Poling Ski | Weak Side Ski | No-Poling Ski |

FIGURE 7.2. Ski skating terminology (Nelson *et al.*, 1986).

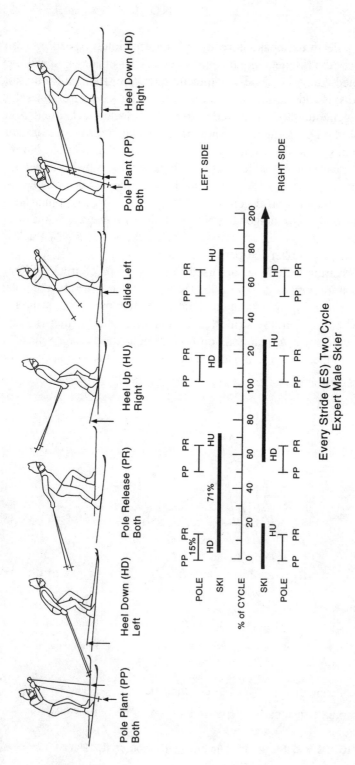

FIGURE 7.3. Every-stride (ES) ski-skating technique. The top panel shows the dwell time for each segment. The lower graph shows the maneuvers for one half cycle. Note that the top panel and the graphs come from two different sources. Although they relate well to each other, the relation is not exact. (Skier images from video of P. Peterson, PSIA Demo Team, used with permission. Data for the graph from Nelson et al., 1986.)

of propulsion. A complete every-stride technique cycle consists of two double-pole plants. The cycle starts with a double-pole plant and a push off from the poles followed by a glide. The skier then propels himself further by pushing off on one ski at an angle of 10°–20° to the line of motion while the poles are simultaneously brought forward for the next double-pole plant and push, after which the skier pushes off of the opposite ski, completing the cycle.

The sequence of the actions shown in Fig. 7.3 is as follows: the skier moves through both pole plants (PP), left heel down (HD), both poles release (PR), right heel up (HU), right side unweighted, the final dual pole plant (PP), and right heel down (HD), which begins the next half-cycle. When both heels are down, the skier is in the glide phase; in the ski recovery phase, one or the other of the skis is unweighted and the skier has that heel up.

The AS technique, shown in Fig. 7.4 is asymmetrical, which makes it somewhat more complicated than the ES technique. On one side of the skier, called the glide or strong side, the pole and ski motions are synchronized. On the other side, the poling and the ski glide are out of phase. Both skis are used for the propulsive skating motion, but the poling and skating motions are performed together on one side only (the strong side). Note that the poling ski is synchronized with the glide pole; the other pole is the nonglide pole and the other ski is the nonpoling ski. The cycle starts with the heel down on the right-hand side, nonpole ski (HD/NP). The nonglide right pole is planted (PP/NP); then the left glide pole is planted (PP/G) and the heel comes down on the left pole ski (HD/P), which is gliding with the same side pole thrust. The heel comes up on the right nonpole ski (HU/NP) at about the same time. The right nonglide pole releases (PR/NG); then the left glide pole releases (PR/G) as the skier pushes off of the glide or pole ski until the heel comes up (HU/P) at about the same time. When the heel is down on the nonpole ski, the cycle is ready to repeat.

Another way of looking at the alternate-stride cycle is to think of the pole and the strong side or pole ski pushing together: the skier thrusts the pole forward for the plant while flexing his knee, and then the skier pushes off simultaneously against the planted pole and the edged ski. The strong side pole releases as the skier pushes off, putting the weight onto the weak side or nonpole ski; neither pole is planted until after the skier pushes off from the weak side ski. Then the skier plants the weak side pole to move his body over to the strong side ski to repeat the process.

The graphed information in Figs. 7.3 and 7.4 show which actions are performed in what sequence and for what percentage of time during a complete cycle. The thin lines represent the pole times and the heavy

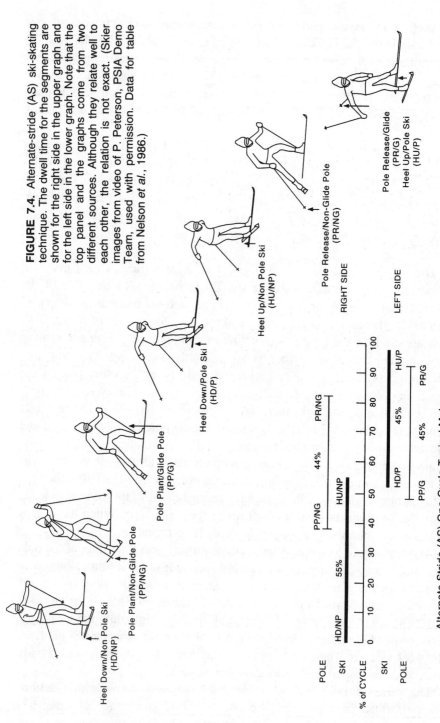

FIGURE 7.4. Alternate-stride (AS) ski-skating technique. The dwell time for the segments are shown for the right side in the upper graph and for the left side in the lower graph. Note that the top panel and the graphs come from two different sources. Although they relate well to each other, the relation is not exact. (Skier images from video of P. Peterson, PSIA Demo Team, used with permission. Data for table from Nelson et al., 1986.)

TABLE 7.2. *Cycle rate, cycle length, and velocity for every-stride (ES) skiers.[a]*

	Top group	Bottom group
Cycle rate (cycles/s)	0.85	0.78
Cycle length (m/cycle)	3.81	3.58
Velocity (m/s)	3.25	2.82

[a]Nelson *et al.*, 1986.

bars are the ski times. The percentages given are for male ski-skaters on moderate slopes. On steeper slopes the dwell times for the skis and poles, that is, the periods in the cycle during which the skis and poles are in contact with the snow, increase somewhat.

The cycle rate, cycle length, and velocity achieved by two groups of competent nordic skiers, one of high proficiency (top group) and the other of lesser proficiency (bottom group), using the every stride technique are listed in Table 7.2. The term *cycle* refers to the skier's stride.

Table 7.3 gives the mean values for the several variables listed that were achieved by nordic skiers using the AS technique.

In general, cycle or stride rates control the velocity of skiers using diagonal stride techniques. The stride length remains relatively constant. The same is true for marathon ski-skaters: stride lengths remain constant while the skiers change their tempos. For ES skating, the cycle rates for faster and slower skiers were comparable. Faster ES skiers had longer cycle lengths. All skiers, however, tended to increase their velocity at any speed by increasing their cycle rates rather than their cycle lengths. In a race, the skier strives for the fastest cycle rate. Only during training would skiers emphasize improving their stride length.

Kinematic analyses of AS ski-skaters are very detailed and complex, and they include summaries of force measurements derived from both snow skiers and roller skiers [2]. These analyses have been done for alternate-side ski skating only. Poling forces are as large as 40–50% of the skier's weight, and they increase with speed. These poling forces are 2–4 times larger than the poling forces generated in diagonal stride skiing. As might be expected,

TABLE 7.3. *Mean values of variables for alternate-stride (AS) skiers.[a]*

Race velocity	6.05 m/s	Strong side ski angle	23.9°
Cycle velocity	3.19 m/s	Weak side ski angle	22.5°
Cycle length	3.83 m	Strong pole phase	0.56 s
Cycle rate	0.83 Hz	Weak pole phase	0.50 s
Strong side ski velocity	3.05 m/s	Strong ski phase	0.56 s
Weak side ski velocity	2.50 m/s	Weak ski phase	0.65 s

[a]Nelson *et al.*, 1986.

the ski provides most of the lateral force component for the side to side action, and it also provides most of the weight support. Propulsion is distributed evenly between the poles and skis. The strong side pole supplies the largest propulsion, followed by the weak side pole, strong side ski, and weak side ski, in that order. The pole forces increase the speed more than the skating forces. Compared with the diagonal stride, the alternate-side ski-skating stride has about 70% greater duration. The peak skating forces are 1.2–1.6 times the skier's body weight, and the peak poling forces are 0.5–0.6 times the body weight. On grades of 9–14%, poling contributes about 66% of the propulsion, but a only a small portion of the lateral and normal forces. Thus in alternate-side ski-skating, the legs support the body and provide the lateral motion while the arms supply the propulsive force. Clearly the 18–20% higher speeds recorded for ski-skating result from the larger average pole propulsion force and the smaller forces that must be applied normal to the ski track for proper traction during the kick phase.

ADVENTURE SKIING IN THE BACKCOUNTRY

Skiing groomed alpine slopes or prepared nordic tracks presents most skiers with a range of thrills and challenges more than adequate to fulfill their needs. For a growing number of skiers, however, the untracked—and often uncertain—snows of the backcountry and remote mountain terrain offer the ultimate adventure in skiing. Adventure skiing often requires some modification of both equipment and technique; certainly it requires that skiers take responsibility for their safety, relying on themselves and, in many cases, their guides to prepare for the hazards that can accompany this special type of skiing.

Off the groomed slopes and prepared tracks, the snow will no longer have a uniform consistency and hardness, the terrain will vary much more greatly, and obstacles will not be marked or, in many cases, even apparent. Consider the predicament faced by the unfortunate skier shown in Fig. 7.5. One of the authors of this book (D.L.) found himself in this situation once and was fortunate to be able to release his free-heel bindings and dig his way out.

The threat of avalanche is probably the most dramatic hazard of adventure skiing in the backcountry, but there are other less dramatic but just as serious situations that the skier may encounter [3]. Our purpose in this chapter is not to prepare readers with everything they need to know for skiing the backcountry, but to offer some understanding of the physical properties of the equipment used in adventure skiing, the backcountry snow itself, and the techniques that skiers will need to use if they choose to seek some adventure by skiing the backcountry slopes.

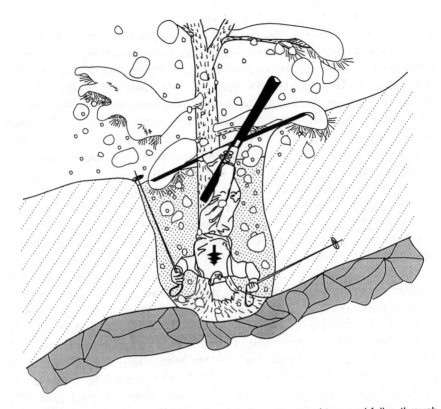

FIGURE 7.5. This skier has skied too close to a snow-covered tree and fallen through the snow to hang upside down in the well that has formed around the tree.

EQUIPMENT

Skiers taken by snow cat or helicopter to remote areas for high-mountain powder skiing often use alpine skis that are usually not longer than 190 cm with a width of sufficient area to support the skier's weight in powder snow. These skis have a low edge profile and are relatively soft in flexure and torsion so that a large reversed curvature results when they are loaded under the skier's weight. Adventure skis should be relatively light and have a low swing weight to permit easy unweighting and rotation. Such skis are easy to turn and readily adaptable to variations in the terrain. Like all alpine skis, they have a single camber.

If the adventure skiers are dropped off at the top and picked up at the bottom of a slope, alpine boots and bindings are a good choice. Adventure skiers who intend to ski up as well as down the slopes of the backcountry use alpine skis like those described above fitted with bindings that can be alternately released at the heel for climbing and fixed at the heel for down-hill use. The free-heel equipment used for nordic touring generally will not work well in the mountainous backcountry. Nordic skis are not wide enough; they lack the flotation area needed to support a skier in deep powder snow. The common, three-pin toepiece boots and bindings used in nordic skiing offer inadequate boot–ski control, and the soft boots do not provide the lateral support needed for running untracked snow. Telemark skis or powder skis will serve, provided they have sturdy, free-heel bindings and the boots used with them have good lateral ankle support.

Finally, skiers who intend to ski up as well as down steep backcountry slopes should not rely entirely on wax for traction. Climbing skins, so called because they originally were made of seal skin, are essential for extensive adventure skiing off well-traveled trails and on steeper pitches. Today's skins are usually made of mohair or nylon fabric with a short nap fitted to cover either the full length of the ski or under the foot area only. The better-quality climbing skins are coated with an adhesive that securely anchors the fabric to the waxed running surface of the ski. Climbing skins are readily removable and reapplied, even in cold weather, but they need to be recoated with adhesive about once a year. Less expensive skins must be fastened to the ski by straps that go around the ski, which can cause prob-lems. These retaining straps can interfere with effective edge control on hard surfaces. In addition to their more obvious importance for uphill travel, skins are very useful going down steep slopes with a loaded pack. The energy saved in controlling speed and in avoiding spills is well worth the trouble to put them on. Furthermore, on a steep slope they permit carving turns down the fall line while carrying a loaded pack. That advantage can decrease the overall travel time markedly, particularly on routes that involve several up and down sections.

SNOW COMPACTION

When skiing in soft, uncompacted snow, the entire bottom running surface of the ski is supported by the snow that the ski compacts beneath it, so turning in powder snow results from the overall flexure of the ski, not from carving the snow with the inside edge. The dynamics of linking turns through deep powder using up–down weighting and body angulation are quite different from the dynamics of linking carved turns on a hard-packed, groomed slope. The groomed slope is very forgiving in ways that untracked powder, off-piste, or crud snow is not.

The properties of the ground-cover snow bed may vary widely depending on the slope aspect, the weather, and the age of the snowpack, making it possibly one of the most complicated viscoelastic media encountered in any engineering application [4]. As we saw in Chap. 2, the snow in the pack consists of rounded ice grains that have been bonded together by vapor condensation at their contact points. Because ice tends to flow when it is loaded, the mechanical behavior of the snow will depend on the viscoelastic behavior of the ice bonds in the snowpack and the relative void and ice-grain composition of the snow. Very low density snow has a large void volume with few ice grains and bonds; such snow will compact readily and suffer a large change of volume when it is loaded. At the opposite extreme, in glacial ice, which was once new-fallen snow, all the voids have closed, and it can be elastically deformed; or, if loaded for sufficient time, it can become permanently deformed.

Aged snow that is capable of supporting a skier presents the strain–time curve shown in Fig. 7.6. As soon as loading is applied to the snowpack, elastic deformation takes place and the strain rises to the first step (1) noted in the figure. If the load is immediately removed, the strain returns to zero. If the load persists, an additional strain appears over time, illustrated as step (2). The strains produced by the loading on the snowpack represented in steps (1) and (2) are reversible: if the loading is removed, the strain disappears, as illustrated in steps (4) and (5). If the load is applied over a long period of time, as illustrated by step (3), the snowpack continues to deform as the ice grains are rearranged and packed to fill the void spaces until the strain finally saturates and only the flow of ice remains. Step (3) represents the progressive compaction of snow. On a groomed slope, skiers have no effect on the progressive compaction of the snow because the grooming process applies much higher loads to the pack than any number of skiers

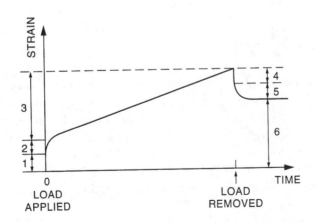

FIGURE 7.6. Strain-time curve for a snowpack (Mellor, 1964).

can. Skiers on groomed slopes either cut or abrade the thin surface layer of the snowpack; they do not compact or otherwise affect to any degree the strong, underlying icy layer of snow.

Rapid loading of a semi-infinite snow mass compacted by a flat plate can be expressed for a range of snow densities and penetrations by the relation

$$p = ky^n, \tag{7.1}$$

where p is the compaction pressure, given by the loading of the flat plate divided by its area; y is the depth of penetration of the plate into the snow mass; and k and n are constants for a range of p and y that depends on the snow type, density, and temperature. The value for n seems to be relatively insensitive to variations in the snow type, decreasing only slightly with snow density; the value for k varies several orders of magnitude with the snow density and depends on the snow type as well as the dimension of the plate. Figure 7.7 uses experimental data gathered from several sources to plot the value of the compaction pressure p against the depth y for natural snows of different densities [5]. Notice that the compaction resistance values vary rapidly with the different values for snow density. The curves for snow densities, ρ of 0.06–0.15 g/cm³ are relatively good matches with the pressures and compaction experienced in backcountry skiing.

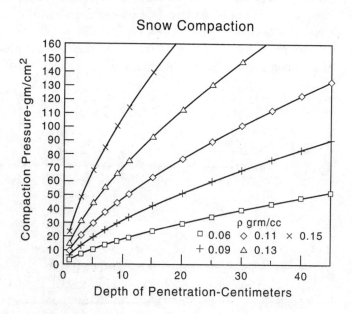

FIGURE 7.7. Snow compaction plots vs depth for snows of different densities, ρ in the range of 0.06–0.13 (data from Mellor, 1964).

DYNAMICS OF CARVED TURNS IN SOFT SNOW

Earlier discussions of carved turns assumed that the ski was carving a groomed slope; that is, the edge of the ski pressed against a completely rigid surface. On a groomed slope, the flexural shape of the turning ski is determined by the ski's geometry, and the distributed loading on the ski can be calculated from its bending and torsional stiffness and the curvature at each segment of the ski. Analyzing turns made in the backcountry snowpack presents a different problem. Once again, the geometric shape of the ski is the relevant factor. But that shape—the reversed camber the ski assumes when it bows under loading as it compacts the uncompacted snow beneath it—depends greatly on the density of the snow and several other parameters. The different roles played by the parameters involved in the reversed camber process are worked out in detail in Technote 12, p. 244, "Ski Flexure in Uncompacted Snow." In the discussion that follows, we consider the geometry of the dynamic process of carving parallel turns in soft snow.

When a skier loads the ski to make a turn in soft snow, the reversed camber that results lifts the ski tip up relative to the track the ski makes through the unpacked snow while the rest of the ski rides directly in the track. The geometric relationship between the amount of tip lift, Δ, the radius of the ski's reversed camber curvature, R, and the contact length l may be expressed as

$$\Delta = \frac{l^2}{2R}. \tag{7.2}$$

With the ski tip riding above the ski's track through the snow, the skier uses body angulation to rotate the ski at an angle relative to the plane of the slope, Ψ. The reversed camber of the curved ski projects an arc on the plane of the snow that varies according to the value of Ψ, and the ski turns through the snow, following the path of that projected arc.

This process is modeled in Fig. 7.8. In panel (a), a skier makes a track through soft snow, and with his ski tips raised above the track he has begun to angulate his body to rotate his skis and make a turn. Note that the tails of his skis run directly in the track made in the snow; for the tails to do otherwise would mean that the skis must plow snow off to the side rather than compact snow beneath them, and the resulting drag force would stop the skier's forward motion. Figures 7.8(b) and 7.8(c) illustrate the geometric relationships associated with the turn the skier shown in Fig. 7.8(a) is about to initiate.

In Figure 7.8(b), the x axis is the direction of motion, the y axis is normal to the slope, R represents the radius of the ski's reversed camber curvature,

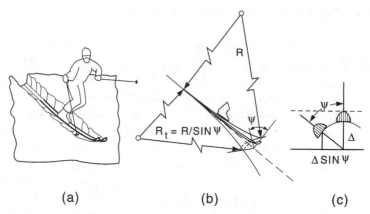

FIGURE 7.8. A skier runs through soft snow (a). Views (b) and (c) illustrate the geometry of the turn the skier in (a) is about to initiate.

and Ψ represents the angle between the ski and the plane of the slope. Figure 7.8(c) offers a frontal view of the ski tip riding above the track in the snow and rotated to make the turn. From the geometry illustrated, we see that the radius of the carved turn that will result, R_t, is equal to the radius of the reversed camber of the ski divided by the value, of the sine of Ψ. If the ski is not rotated at an angle relative to the slope, Ψ is zero, R_t is infinite, and the ski does not turn. If $\Psi = 90°$, the radius of the turn becomes the full value of the ski's reversed camber radius R. For moderate angles between the ski tip and the plane of the slope, the turn radius is determined by the tip-lift projection on the plane of the slope, $\Delta \sin \Psi$, as illustrated in Fig. 7.8(c). The radius of the carved turn in soft snow becomes

$$R_t = \frac{l^2}{2\Delta \sin \Psi} \rightarrow R_t = \frac{R}{\sin \Psi}. \tag{7.3}$$

This description of the turning process in soft snow is based on an ideal model of the ski's geometry as it flexes and assumes a reversed camber configuration.

In the backcountry, skis usually track from at least 10 cm to as much as 40 cm deep in uncompacted snow. Figure 7.9 below illustrates a generic, 190-cm ski running through powder snow with the load placed on the ski in three different ways. In Fig. 7.9(a), the load is distributed uniformly over the ski, which is tilted backward so the tail runs tangent to the ski's track in the snow. In Fig. 7.9(b) the tail of the ski carries the full load; the ski's forebody acts like a paddle deflecting the snow in its path. In Fig. 7.9(c) the ski's forebody carries all of the load; the afterbody just trails behind it in the

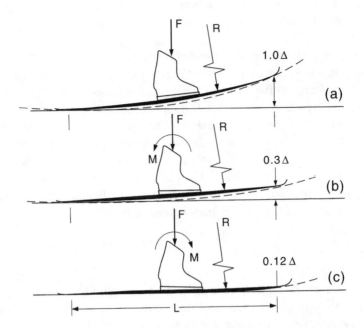

FIGURE 7.9. Limiting cases for generating reversed camber in soft snow for the same ski under different loading.

track. In each case, the radius of the ski's reversed camber curvature, R, is estimated by passing the arc of a circle through the tip and tail of the ski. In actual practice, a ski's reversed camber radius will lie somewhere between the extreme cases illustrated by this model.

The turning radii R, for the cases illustrated above may be obtained using the formula given in Eq. (7.3). The contact length l for a 190-cm ski is 168.5 cm and the tip lift value $\Delta = 16.4$ cm, so the ski's reversed camber radius R is 8.7 m with an evenly distributed load, as shown in Fig. 7.9(a), and the ski turns through the snow effectively. When the afterbody of the ski is fully loaded and the ski's forebody is straight, as it is in Fig. 7.9(b), the reversed camber radius becomes 29 m, which makes the ski much more difficult to turn. To achieve this afterbody loading, the skier would have to apply a backward torque at the boot to put his center of mass directly over the afterbody of the skis. When only the forebody of the ski is flexed, as it is in Fig. 7.9(c), the reversed camber radius of the ski becomes 73 m, which would make turning the skis in soft snow essentially impossible. Note that the example shown in Fig. 7.9(c) is given solely to illustrate how the turning radius changes when the ski flexes under a load. An actual skier loading the forebody of the ski in the manner shown in Fig. 7.9(c) would have to apply a forward torque to the ski at the boot, which, except in very strong snow, would probably cause the skier to pitch forward and fall because the skis would dive into the snow.

Let us now compare carving parallel turns in soft snow with the telemark turn, which has some of the characteristics of both a carved turn and a wedge turn. Refer to Fig. 4.14, p. 107, which shows the wedge configuration assumed by skis in a telemark turning stance having an angular opening of θ. Assume that the ski's contact length is some value l and that the forward ski is shuffled ahead of the trailing ski by one-half of its contact length. Consider the angled, two-ski configuration of the telemark turn as a curved stick that lies parallel to the plane of the slope. In this case, the equivalent of the projected tip lift Δ for this telemark turn becomes $0.75l \sin \theta$, and the turn radius may be found by

$$R_t = \frac{l^2}{2\Delta} \rightarrow R_t = \frac{l}{1.5 \sin \theta}. \tag{7.4}$$

Using our model 190-cm ski to make a telemark turn for an angular opening between the skis of $\theta = 15°$, the turn radius R_t becomes 4.35 m, which allows the telemarker to make a very tight turn.

The telemark turn in soft snow is generated by the combined action of three different processes: the wedging action produced by the angle between the skis, the reversed camber of the lead ski as it tracks through the snow, and the tilt of the skier's body that is necessary to maintain lateral stability. As the skier's velocity increases, the plowing action from the ski's lifted forebody and tip causes the ski to flex more than it does at slower speeds and that flexure generates additional reversed camber that shortens the turn radius. Recall how on a groomed slope a dynamic instability may occur in the latter part of a carved, parallel turn. When the skier tilts his body to regain lateral stability, he shortens the radius of the turn and increases the outward centrifugal loading. In soft snow, both alpine and telemark turns generate the same radius shortening due to the increased flexure of the ski. The dynamic instability caused by the resulting increase in centrifugal force at the completion of the turn, particularly when making a telemark turn, propels the skier downslope and sideways over his skis as he completes his turn.

Let us look briefly at another alternative to making a turn in soft, crud, or crusty snow: the powered or jump turn. When making any kind of turn in soft snow, the skier must use some enhanced up motion to lift the skis so that they may be rotated in the direction of the turn. As we have seen, sitting back on the skis helps the tips ride up so that they break the surface of the snow. With the ski tips free of the snow, the skier can plant his poles and then push up and off with a torquing action that rotates the skis into their new direction as they come free of the snow. This action is certainly necessary when skiing the backcountry on a slope covered with snow that has a breakable crust. The powered turn performed in this manner and under these conditions is a sudden change of direction in which the skier plants a pole, up-unweights her skis, rotates them, angulates her body, and then returns to the snow, now traveling in the new direction.

Look at the photograph in Fig. 7.10, which shows ski tracks made in backcountry powder. The tracks on the left were made by a skier using telemark turns to descend the slope. The tracks on the right were made by an alpine skier using powered turns. Note that the tracks made by the telemark skiers are wider, their curve radius is larger, and the turns themselves suggest a more leisurely, smoother descent of the slope. In contrast, the tracks on the right have much tighter turns with abrupt changes in direction from curve to curve as the skiers powered their way down the slope. Note also the spray of snow debris at the moment when the skier's direction changed as the skis were powered up and out of the snow, rotated, and then reset in their new path. Here we have visual evidence that there is more than one way to descend the ungroomed slopes of the backcountry.

FIGURE 7.10. Four tracks in backcountry powder. The tracks on the left were made by skiers executing telemark turns. On the right, the skiers used powered parallel turns. (Photograph taken by Kerry Walton and reprinted from English, 1984.)

THE PHYSICS OF SURVIVAL

Some of the physics of backcountry, adventure skiing has nothing to do with the physics of ski equipment but everything to do with basic survival. On a backcountry tour, the members of a ski party must always be prepared to survive the night, even if they are no more than a mile or so from a road. Winter survival at night in the backcountry requires carrying proper supplies; chief among them are a small shovel or other digging tool and some kind of a blanket or tarp, preferably one with a reflective aluminum coating. When wrapped around a person, a reflective blanket's coating helps prevent heat radiation loss from cooling the body and creates a dead air space around the body that provides thermal insulation. A tarp can offer some shelter from the elements, but better shelter may be found at the bottom of the snowpack next to the ground where the temperature will usually be exactly at 0 °C or 32 °F. In subartic zones, the ground is heated by conduction from the heat inside the earth, and the temperature at the surface of the ground rises to the melting point of the snow. Such ground melting goes on all winter. The covering snowpack acts as an insulator, holding heat in at the ground level. If the melting point of snow does not seem an especially warm temperature, remember that temperatures at night at altitude in the winter may go down well into double digits below 0 °F, which makes a positive 32 °F a very welcome, relatively quite warm, temperature.

For the best shelter, dig a snow pit all the way down to the ground so that your party can, if possible, sit on the ground itself. Cover the snow pit with tree boughs or with skis, cover that with a tarp, and then sit on the ground wrapped in your reflective blankets. If time, energy, and a shovel are available, a large snow pit with a small fire at one side near the ground will do wonders for a party's morale as they settle in for an unplanned overnight stay. Given the insulating properties of snow, even a candle will heat a small snow cave appreciably.

Snow is the blackest substance known to man in the infrared region of the spectrum. On a clear night at altitude in the winter backcountry, the snow is like a black stove that radiates heat to the emptiness of the night sky where the temperature beyond the atmosphere is −270 °C. Thus the relatively warm snow radiates its heat upward, and the snow's surface layer cools rapidly. When clouds are present, the cloud cover acts as nature's reflective blanket, radiating heat back to the ground. The snow's surface does not cool to the same degree on a cloudy night. Another consideration is that the night air, as it is cooled by the surface of the snow, becomes heavier and drifts down to the valley bottom. Overnight, mountain valleys may become 20 °C colder than the upper mountainsides that surround them. For this reason it is important to stay sheltered from the night breeze of cold air that descends the slopes to the valleys below.

Another aspect of the physical reflective properties of snow can be useful to the backcountry skier. By diffuse reflection from the ice grains, snow reflects up to 95% of the visible light that falls on it. In direct sunlight, an untracked snow surface appears brighter than the pock marks or ski tracks made in the snow by passing skiers. Light falling on a depressed track in the snow does not escape—or reflect—as readily as light that falls on the smooth surface of the untracked snow, which makes the track appear darker. However, in the indirect, diffused light of an overcast day, exactly the opposite is true. The photograph in Fig. 7.11 shows a ski track in the dull light of an overcast day. The depressed track is clearly brighter than the surrounding snow. On a cloudy day the surface of the snow is illuminated in all directions by sunlight that has been scattered in its passage through the overhead clouds. The scattered light is scattered still more by the snow

FIGURE 7.11. Ski tracks illuminated by the diffused light of an overcast day show more brightly than the surrounding untracked snow.

below the surface, and some of that light emerges from the sidewalls of the ski track to illuminate the bottom of the track, making it brighter than the surrounding snow surface.

On a cloudy day, this observation can help a skier detect old tracks in the snow, which are often covered with some new snow. If they are not buried too deeply, the old ski tracks may call attention to themselves by shining up through the snow. These old tracks make for much easier travel because they are already compacted and, having supported a skier in the past, they can support still another skier without further compaction. Not having to break a virgin track in soft or crusty snow saves a lot of energy. Even in the dark, a skier who has just lost his way off of an old track may find his way back by feeling for the hardening in the snow caused by the compaction of touring skis.

There is a great deal of physics involved in understanding and assessing the snow avalanche hazards that attend backcountry travel in a variety of different snow conditions. That topic is complicated enough and important enough that we will not deal with it here because any cursory treatment of the threat posed by avalanches may give readers a false sense of security. Where avalanche danger is involved, a little knowledge can, indeed, be a very dangerous thing. Before skiing into the backcountry in search of adventure, learn about avalanches by consulting one or more of the works listed in the bibliography, or, better, ski with an experienced guide who knows the territory.

REFERENCES

1. See R. C. Nelson, J. McNitt-Gray, and G. Smith, "Biomechanical Analysis of Skating Technique in Cross Country Skiing," Final Report to the U.S. Olympic Committee, Colorado Springs, CO, 1986.
2. See G. A. Smith, "Biomechanics of Crosscountry Skiing," in Sports Med. 9(5), 273 (1990); and E. C. Frederick and G. M. Street, "Nordic Ski Racing, Biomechanical and Technical Improvements in Cross-Country Skiing," Sci. Amer. 258(2), T20 (1988).
3. For information regarding backcountry skiing and the specific hazards presented by snow avalanches, see the several references cited in the bibliographical essay under the heading "Back Country Skiing."
4. For more information on backcountry snow and the snowpack, see relevant chapters in D. M. Gray and D. H. Male, editors, The Handbook of Snow (Pergamon, Toronto, 1981); see also two papers by M. Mellor, "Properties of Snow," Monograph III-A1 (U.S. Army Corps of Engineers, Cold Regions Research and Engineering Laboratory, Hanover, NH, 1964); and "Engineering Properties of Snow," J. Glaciol. 19(81), 15 (1977).
5. The data plotted in the graph come from Mellor's two papers cited in Ref. 4.

CHAPTER 8

FRICTION
Glide and Grab

The study of skis as sliders on snow has a long history related to the more general study of friction as it affects the grinding and abrasion of moving machine parts and the most effective means of lubricating those parts. Skiing is unique among these studies because, while alpine skiers seek minimal sliding friction to increase their downhill speed, track and cross-country skiers require their skis to combine minimum and maximum friction. The running surfaces of track and cross-country skis must offer minimum friction on the forward glide and maximum tractional friction, or grab, when the skier pushes off against the snow for acceleration and for travel uphill. This dual, contradictory requirement makes the problem of analyzing skis as sliders on snow, and the many waxes and running surface configurations used to satisfy the need for glide and grab, very subtle. There is no definitive, scientific analysis of this problem. At best, scientific research can indicate pragmatic solutions to problems whose effectiveness must be tested in the field [1].

In general, we use the term *friction* to classify all of the effects that, taken together, impede the motion of the skier. The simple, mechanical concept of friction as it affects a skier moving at a very low speed is illustrated in Fig. 8.1. The friction force F_f resists the motion of the skis as they glide over a plane surface. The friction force is proportional to F_N, the normal force (in this case, the weight of the skier). The friction force is expressed by the relation $F_f = \mu F_N$, in which μ, the coefficient of friction, accounts for the details involved in the friction process, and those details are many and complicated. For example, consider that the characteristics of a snowpack—its temperature, density, strength, liquid-water content, and snow-grain type, to name only a few—may all affect the coefficient of

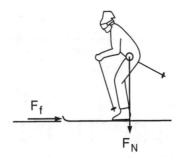

FIGURE 8.1. Friction and the skier viewed as a simple, mechanical concept.

friction. A major reason that alpine slopes are groomed is to reduce the natural variability of the snowpack so that the snow will ski with a degree of uniformity. The condition of a given snow surface is difficult to measure precisely even in a steady state, and when we consider how snow changes when just a single ski passes over it, we get some sense of the many variables that contribute to the complexity of this problem. In the discussions that follow, we consider some of the more interesting processes associated with the friction force and skiing.

SNOW COMPACTION AND PLOWING

When skis run in unpacked snow, the work done to move the skis forward must equal the work done to compact the snow beneath them. In Fig. 8.2, a slowly moving skier compacts an unpacked snowpack by the height h over a forward distance of one ski length l. The depth of the snow compaction is proportional to the weight of the skier, the normal force, F_N; thus

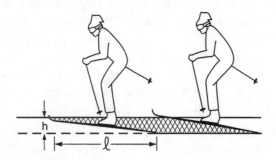

FIGURE 8.2. A slowly moving skier compacts an unpacked snowpack by the height h over a forward distance of one ski length l.

$$F_{comp}l = F_N h. \tag{8.1}$$

The work done by the component of force necessary to propel the skier through the snow the distance l against the compaction force F_{comp} is equal to the work done by the weight of the skier, F_N, in compacting the snow to the depth h. The left-hand side of this equation represents the work done by an external force acting over the distance l. The right-hand side of the equation is not precise because variations in the composition of the snow-pack affect the exact compaction force required, but in every case, the compaction force required will be proportional to $F_N h$. When we cast this equation in the form of a coefficient of friction, we get

$$\mu_c = F_{comp}/F_N = h/l. \tag{8.2}$$

Thus the friction caused by the ski's compaction of the snow equals the coefficient of friction multiplied by the skier's weight. Note that because the ski tilts upward in its track as it compacts the snow beneath it, the compaction force makes the skier, in effect, ski uphill. When skiing in very soft snow, the compaction force may be the dominant component of the friction force. In fact, a skier may sink so far into the unpacked snow that he comes to a complete stop when the upward slope of his own track levels the downward slope of the hill. On a groomed slope, of course, the compaction force illustrated here is negligible.

In addition to the force required to compact snow under the ski, force is required to plow the snow from the ski's path. This plowing force represents a dynamic compaction of the snow that results from the transfer of momentum that occurs between the moving ski and the snow at rest. The plowing force does not, like the compaction force, depend upon the weight of the skier. The plowing force is analogous to aerodynamic drag, so readers interested in a more technical consideration of this subject should see Technote 8, p. 222, "Aerodynamic Drag." The plowing force depends on the frontal area of the ski, the density of the snow being plowed, and, in most cases, on the square of the skier's velocity.

DRY FRICTION

At very cold temperatures, below about $-25\ °C$, the sliding friction of skis on snow changes markedly—the skis grab the cold, dry snow and simply do not slide with anything approaching the ease with which they would slide over the same surface at a higher temperature. Extremely cold temperatures keep the snow from melting under the skis as they pass over the snow surface. Without meltwater for lubrication, the hard, dry ski bottom rasps against the hard, dry asperites of the ice grains in the snow-pack. The qualitative effect of the dry friction that results from skiing on very cold snow is similar to skiing on a sand dune.

To develop an analysis of this process, let us assume that the ice grains in the snow surface fracture and yield under the rapid loading that occurs as a ski passes over it [2]. The total area of the ski bottom that makes contact with these ice grains, A_c, is given by

$$A_c = F_N / \sigma \qquad (8.3)$$

where σ is the unconfined, compressive strength of the ice and F_N is the normal force, the weight of the skier. We assume that dry friction results from the shear fracture of the asperites over the contact area of the ski bottom, A_c. The total dry friction force f_d, becomes

$$f_d = \tau A_c = F_N \tau / \sigma$$

or $\qquad\qquad\qquad\qquad\qquad\qquad\qquad\qquad\qquad\qquad (8.4)$

$$\mu_{\mathrm{dry}} = \tau / \sigma$$

in which τ is the shear strength of ice. Using measured values of the tensile and compressive failure stresses for ice, we find that the coefficient of friction for dry snow, μ_{dry}, drops from 0.32 at 0 °C to about 0.13 at −32 °C [3]. These values are much larger than those observed using actual skis in the field. The process of dry friction plays very little role in the enhancement of a ski's glide. It is more important, as we shall see, in the processes that increase a ski's ability to grab the snow.

In our discussion, we have assumed that the ski bottom is much harder than the ice. If it were not, the fractured ice grains would likely embed in the ski bottom and shear away from the snow surface; snow and ice would build up on the bottom of the ski and further decrease the ski's ability to glide. When skiing at very low temperatures, skiers may avoid this effect by preparing the running surfaces of their skis with waxes that create the hardest, smoothest surface possible, the base preparation that gives the best glide in such very cold conditions.

MELTWATER LUBRICATION AND VISCOUS FRICTION

The presence of meltwater lubricating the running surface of the ski has been confirmed and measured using a probe on the bottom of a ski; the thickness of the water film varied according to the type of wax preparation used, the temperature of the snow surface and the ski bottom, and the velocity [4]. Other measurements made of the temperatures found on the running surfaces of alpine skis suggest that liquid water must be present at the interface of the skis and the snow [5].

Additionally, much of the general research on slider friction supports the idea that meltwater provides the lubrication that helps skis slide over the snow. We can test this idea by using known values for the coefficient of friction to calculate the work done to push a ski over the snow. We then assume that, in the steady state, all of the work done to push the ski over the snow produces frictional heating at the interface of the ski bottom and the snow that goes into melting the ice grains in the snow surface to form the lubricating layer of meltwater. For a more technical discussion of the friction processes that create the meltwater film on the running surface of the ski, see Technote 13, p. 249, "Meltwater Lubrication."

If we assume that the meltwater is a thin film of liquid that completely separates the ski base and the ice grains, then the viscosity of the liquid film must be responsible for the heating that occurs. The viscous friction force is given by the relation

$$f = \eta A v / h, \tag{8.5}$$

where η is the viscosity, A is the area of the liquid film, v is the velocity, and h is the thickness of the liquid film. Model studies of the creation of a lubricating film of meltwater on the running surface of a ski uniformly lead to values for the liquid film that are so thin that the asperites on the prepared ski bottom must impinge on the ice asperites in the surface of the snow. This action results in more than a simple rubbing and fracture process because both the ski bottom with its wax preparation and the ice grains in the snow surface suffer plastic deformation; that is, their shape changes and stays changed. Figure 8.3 shows four possible interactions that may occur at the interface of the ski bottom and the surface of the snow.

In Fig. 8.3(a), a polished, hard ski base deforms elastically as it passes over a surface composed of old snow in which the ice grains have been smoothed by melting at the high points and refreezing in the valleys. In Fig. 8.3(b), a hard ski base polishes the sharp ice asperites present in a surface composed of dry, new snow by fracturing them. In Fig. 8.4(c), a soft ski base slides on old snow lubricated by a liquid film. Note that the meltwater does not form a continuous film over the entire ski base. Finally, in Fig. 8.3(d) a soft ski base is abraded by the asperites of dry, new snow grains.

The least friction, and thus the best continuous glide, is obtained by the combination of conditions shown in Fig. 8.3(c). The conditions shown in panels (a) and (b) have greater friction than (c), and they would result in some combination of glide and grab. The combination shown in (d) results

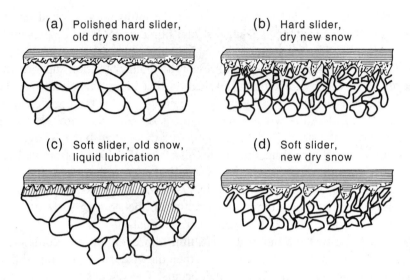

FIGURE 8.3. Four examples of the interactions that may occur at the interface of the ski bottom and the surface of the snow.

in the greatest friction among the conditions shown and thus the greatest grab and the least glide. The combination shown in (d) would likely be useful only for skiing up a slope. After reaching the top and before going back down, assuming that the snow conditions and the temperature remained roughly the same, the ski base would likely be scraped clean and prepared using a different, probably harder, wax if the skier wished to glide down the same hill that her skis, by grabbing the snow surface, had just helped her climb up. Given the wide range of temperatures and snow conditions a skier may encounter on any given day, both the importance and the complexity of using the proper wax on the proper snow surface at the proper temperature to prepare the ski so that it glides and grabs the snow in the desired manner should be apparent.

CAPILLARY AND TRIBOELECTRIC DRAG EFFECTS

Looking again at Fig. 8.3, note that in each of the cases illustrated there are many ice grains in the snow surface that do not come in contact with the running surface of the ski, which means that the spaces between those ice grains may serve as pores in the snow surface through which capillary columns of meltwater connect to the ski base and produce a drag effect on the ski. Laboratory demonstrations using ski base materials and water suggest that such a capillary drag effect exists [6]. Figure 8.4

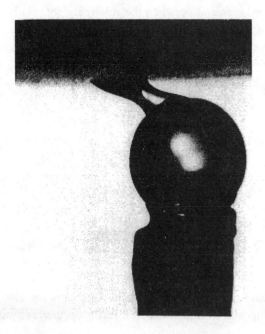

FIGURE 8.4. Capillary attachment between a glass bead and the dry surface of a block of material used to manufacture ski bases (Colbeck, 1992).

shows a capillary column of water connecting a glass bead and the surface of a block of material used to manufacture ski bases that functions here as a representative slider. The slider in the illustration is moving from right to left. Notice the difference between the contact angles that form on the upstream and downstream sides of the capillary column of water where it attaches to the slider. Note also how the capillary column of water stretches as the slider passes over the bead. The change in the contact angle on the leading and trailing sides of the capillary column creates additional drag on the slider, as does the force required to stretch the capillary column of water. This effect explains in part why, when we encounter a wet spot in the snowpack, our skis noticeably slow. In really wet, spring-skiing conditions, the snow has a very high meltwater content. Such wet snow may seem to grab the skis as capillary drag tugs at their running surfaces.

Triboelectric effects may also contribute to the drag on a ski. When a ski rubs over the surface of the snow, it becomes electrostatically charged in a manner similar to the way a glass rod becomes charged when it is rubbed with a cloth. If the snow surface contains any accumulation of dirt or rock dust, the electrostatically charged ski base will attract that dust. The presence of such hard material between the relatively soft ski base material and the surface of the snow leads to increased drag on and abrasion of both surfaces.

EMPIRICAL MODELS OF SKI FRICTION

Even the most carefully formulated empirical models of the friction processes involving skis sliding on snow do not do a very good job of describing the actual performance of skis in the field because of the many variables encountered on the slope. Different variables, such as temperature, the type of ice crystals in the snowpack, the type of wax on the running surface of the ski, and the skier's velocity, dominate the friction process under different conditions. For example, at very cold temperatures with a snowpack consisting of fine-grained ice crystals, the friction process at low speeds is largely determined by the fracturing of ice-crystal asperites under the running surface of the ski. However, for any combination of conditions we might propose, the most important parameter is the effective thickness of the lubricating film of meltwater that forms on the running surface of the skis.

Figure 8.5 presents qualitative values for the contributions made to the total coefficient of friction (μ) by meltwater lubrication (μ_{lub}), dry friction (μ_{dry}), and capillary drag (μ_{cap}) considered relative to the thickness of the lubricating film of meltwater that forms on the running surface of a ski. This

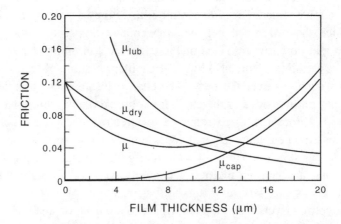

FIGURE 8.5. Qualitative values for the contributions made to the total coefficient of friction (μ) by meltwater lubrication (μ_{lub}), dry friction (μ_{dry}), and capillary drag (μ_{cap}) considered relative to the thickness of the lubricating film of meltwater that forms on the running surface of a ski (Colbeck, 1992).

simplified, qualitative model gives us a good sense of the interaction of these three friction domains and how they are controlled by the thickness of the meltwater film, which, therefore, is used as the independent variable. Note that meltwater lubrication begins to decrease the friction on the ski bottom (which initially is entirely due to dry friction) when the film is just under 4 μm thick and that capillary drag begins to increase the friction on the ski bottom at just after that point. The capillary drag effect exceeds the meltwater lubrication effect when the meltwater film becomes greater than 15 μm thick.

WAXING FOR GLIDE AND GRAB

There is very little information available on the chemistry of ski waxes, and there is almost nothing in the technical literature regarding the scientific basis that underlies the waxing techniques that have been developed over many years of practice applying ski waxes under different conditions to achieve particular results [7]. If such scientific information does exist, it is probably in the proprietary files of ski-wax manufacturers.

Using wax to prepare the running surface of a ski seals the ski base and creates a suitable surface to which subsequent layers of more specialized running waxes can adhere. All modern alpine ski-base materials are made from sintered polyethylene and thus are porous, so the initial "hot wax" base preparation consists of a very hard wax that is melted and spread over the ski base (often with a hot iron) so that the wax thoroughly impregnates the running surface. Then the ski base is buffed to create a polished, smooth surface. Any subsequent applications of running wax are laid down on the hot waxed base in thin layers that are between 0.005 and 0.02 mm in thickness. If these final wax layers were any thicker, they would be more likely to pick up dirt from the surface of the snow, which, as we have seen, would increase friction.

Ski waxes consist of straight-chain or branched hydrocarbon molecules derived from petroleum (crude oil) or coal tar sources [8]. Figure 8.6 shows models of the molecular structure of a straight chain (a) and a branched (b) hydrocarbon molecule. These hydrocarbons are used in the manufacture of all ski waxes.

The straight-chain hydrocarbon molecule $C_{20}H_{42}$, is a low-melting-point paraffin wax that would be used to manufacture a glide wax designed to be applied at relatively warm temperatures. The straight-chain hydrocarbon molecules found in ski waxes range in length from 18 to 50 carbon atoms, that is, from C_{18} to C_{50}; the more carbon atoms there are in the chain, the higher the molecular weight of the molecule. These long molecules orient

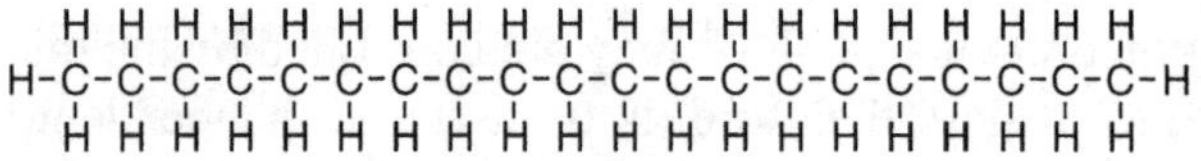

Typical Straight Chain Hydrocarbon Molecule in Ski Wax ($C_{20}H_{42}$)

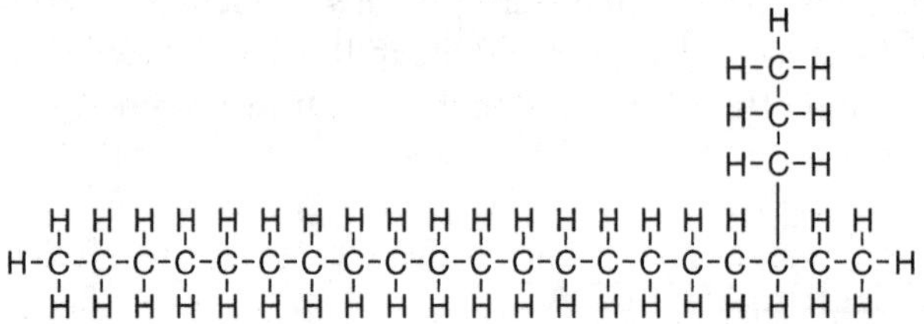

(a)

Wax Molecule with a Single Branch ($C_{23}H_{48}$)

(b)

FIGURE 8.6. Molecular structure of (a) a straight-chain and (b) a branched hydrocarbon wax molecule.

themselves in parallel arrays that maximize their intermolecular bonding. The resulting strength of the intermolecular bonding gives the waxes made with these molecules relatively higher melting temperatures (that increase with increased molecular weight), which makes the waxes harder at snow temperatures and thus more suitable for formulating glide waxes.

Branched hydrocarbon molecules may have one or more short hydrocarbon chains branching off of the primary chain. Because of their branched molecular structure, the intermolecular bond for branched hydrocarbon molecules is not as strong as it is for the straight-chain molecules, so waxes made with branched hydrocarbon molecules are softer and have lower melting points, making them more suitable for the manufacture of waxes designed to increase traction or grab.

The basic constituents of all ski waxes are petrolatum (petroleum jelly); varieties of paraffin waxes having low, medium, and high melting points; microwax; and Fischer Tropsch wax. Petrolatum consists of low-molecular-weight branched hydrocarbon molecules, which makes it a relatively soft substance with a low melting point. Petrolatum gives ski waxes an oily feel and lowers their hardness. Paraffin waxes have a high proportion of straight-chain hydrocarbon molecules with a range of molecular weights. This quality makes paraffins relatively hard substances that exhibit a variety of melting points. Low-melting-point paraffins have molecular weights that range from C_{20} to C_{28}; medium-melting-point paraffins have molecular weights that range from C_{22} to C_{32}; and high-melting-point paraffins have molecular weights that range from C_{25} to C_{35}. Low-melting-point paraffins are major constituents in all alpine and cross-country glide wax preparations. Microwax consists of branched hydrocarbon molecules that have very high molecular weights, ranging from C_{28} to C_{50}. Unlike other waxes composed of branched hydrocarbon molecules, microwax is extremely hard and has a high melting point. Microwax is even significantly harder and has a higher melting point than any of the paraffins. When mixed with the other hydrocarbon materials, microwax raises the melting point and hardness of the composite wax while allowing the wax to remain relatively pliable, probably because of microwax's branched molecular structure. Finally, unlike all of the other waxes, Fischer Tropsch wax is derived from coal tar, not from crude oil (petroleum). It has an extremely high molecular weight and is the hardest primary ski-wax constituent. All of these raw wax products are readily available commercially, and ski-wax manufacturers use them as the basis for all of their various wax formulations.

Base Preparation and Waxing for Good Glide

When laboratory formulations of composite waxes were made to simulate the properties of some commonly used commercial glide waxes, researchers found that only three raw wax products were required: petrolatum, low-melt (LM) paraffin, and microwax [9]. The horizontal bar graph in Fig. 8.7 illustrates the relative composition of several commercial ski waxes formulated to promote good glide. The color of a ski wax is entirely the result of artificial dying. The different colors indicate the snow and temperature conditions that the different waxes are formulated to address. For example, a red glide wax is softer and should be used to promote glide on warmer, wetter snow; a green or blue glide wax is harder and should be used to promote glide on colder, drier snow. But a quick look at the illustration shows that in some cases commercial waxes with different colors have essentially the same constituent composition, despite having different ranges of recommended application. For example, Toko alpine green and glider red are constituently the same wax.

It appears that, at least for glide waxes, fewer types of wax are needed to address the range of commonly encountered snow conditions and temperatures. The underlying physical reality that supports this conclusion derives from the two properties of interest for any ski wax: its hardness and its melting point. One way to measure of the relative hardness of different substances is to compare their homologous temperatures. The homologous temperature is the ratio of a substance's absolute melting temperature T_{melt} to its absolute temperature T. Absolute temperatures are expressed in kelvins; recall that 0 °C equals 273 on the kelvin scale. Thus the homolo-

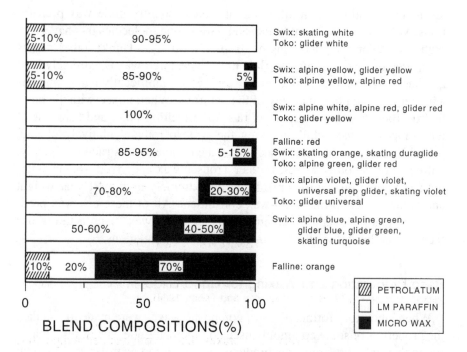

FIGURE 8.7. Composition of several commercial glide waxes in terms of their three principal constituents: petrolatum, low-melt (LM) paraffin, and microwax (Street and Tsui, 1987).

gous temperature of a substance is $T_{hom}=T/T_{melt}$. All glide waxes have melting temperatures that range from about 50 to 100 °C, so their T_{hom} values at 0 °C range from 273/(50+273), or 0.85, to 273/(100+273), or 0.73. The ice crystals that make up snowpacks that have temperatures ranging from about −20 to 0 °C would have effective T_{hom} values ranging from about 0.93 to 1.0, respectively. Within this range, which is roughly the temperature range we encounter when we ski, the physical properties of ski waxes (especially their hardness) change much less with variations in the temperature than do the physical properties of the ice crystals that make up the snowpack.

Figure 8.8 illustrates how the hardness (the force per unit area required to deform the material) of some common ski-base materials and ski waxes stays relatively unchanged as the temperature rises from −20 to 0 °C. Over that same temperature range, the hardness of ice decreases dramatically as the temperature nears 0 °C. Because hardness is the principal effective characteristic of a glide wax, these facts suggest that fewer waxes may be needed to get good glide under most snow and temperature conditions.

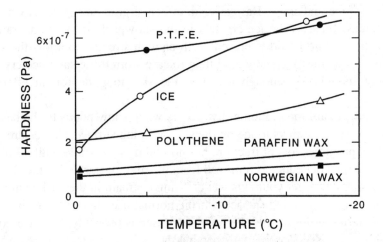

FIGURE 8.8. Hardness of ski-base materials [Teflon (P.T.E.F) and polyethylene], ski waxes, and ice vs temperature (Bowden and Tabor, 1964).

Developers of commercial ski waxes, perhaps responding to these indications, have begun to market glide waxes that have expanded temperature ranges of application. When we want good glide, we need perhaps two or three different waxes, chosen for their relative hardnesses, not the half dozen or more varieties a skier might have chosen from in the past.

Finally, note that although some Teflon-based glide preparations are used for cross-country racing, they are not very resistant to abrasion (the P.T.E.F. line crosses the ice line on the graph at about −12.5 °C). These preparations are very expensive, and they can survive no more than one run over the average nordic race track. It is doubtful that they would survive for even half of one run at an alpine event held on the extremely hard courses prepared for alpine racing.

Base Preparation and Waxing for Traction

Nordic skiers who ski cross country or on a prepared track and adventure skiers who often ski up the slopes they ski down face a very different challenge when they wax the running surfaces of their skis: their skis must alternately grab the surface of the snow and then glide over it. There has been no systematic scientific analysis, such as we referred to above when we considered the friction processes associated with enhanced glide, that measures or models the friction processes associated with the alternating glide-and-grab cycle that occurs in nordic and adventure skiing. In the discussion that follows, we offer plausible arguments that derive mostly from our personal field experience and secondarily from references to the physics of ice and snow. Basically, an effective grab-and-glide wax preparation must bond to the ice crystals in the snowpack at low temperatures where little or no liquid is present with enough strength to allow the skier to push off against the snow, and then release from those same ice crystals quickly and cleanly enough to allow the skis to glide for an effective distance.

First, disregard the grabbing that occurs when snow bonds to the base of the ski because it sticks to ice on the ski's running surface. To be sure, this situation promotes lots of traction, but no gliding is possible until all of the snow and the bonded ice is removed from the running surface of the ski. For this reason, never get your skis wet crossing a stream in the backcountry. If ice forms on the running surfaces during normal use, then either or both the base material of the ski and the wax being used is probably, for a variety of possible reasons, not properly dispersing the meltwater film that, as we have seen, forms on the running surface of the ski. The thin film of ice that results will quickly pick up snow and defeat all efforts to get an effective glide. If this continues to happen even after completely cleaning the ice from the ski's running surface, scrape the ski bottom to remove the topmost layer of wax and apply a harder wax that is likely to be more water-repellent.

The basic phenomenon that produces the grab-and-glide bond must involve the embedding of snow grains into the wax that makes up the ski-base preparation. As long as the skier applies force to the ski so that it presses against the snow, the strengths of both ice and wax are sufficient to prevent either material from suffering shear failure, so traction is possible. As soon as the skier removes the pressure of the ski against the snow, as long as no molecular bonding has occurred (which would happen if the snow were to bond to a layer of ice on the running surface), light abrasion of the ski's running surface against the snowpack allows the snow to be rubbed away as the glide phase ensues. When the ski's running surface is set into motion, the ski achieves good glide over the snow until the forward motion ceases.

During the grab phase of the cycle, the wax on the ski's running surface must deform in milliseconds to allow the snow grains to embed enough to anchor the base of the ski. This process may be readily observed by skiing on frozen, melt–freeze spring snow using a very soft, klister ski wax as a base preparation. After only a few grab-and-glide cycles, the running surface of the ski base becomes pock-marked with ice grains from the snowpack, and the ski will not glide well until the pocking is polished out by the same ice grains in the snowpack abrading against the wax during the glide phase of the cycle. We may also observe how the grab process must work when we climb a packed trail on skis and experience only marginal traction. By stepping off to the side of the trail where fresh surface snow or hoar ice crystals are present (both of which make better asperites than the rounded snow grains found in a packed trail), we find that our skis grab and provide better traction, even without changing waxes. Another maneuver that increases traction is to slap the ski down on each step. The added pressure placed on the running surface of the ski enhances the embedding of the snow grains into the wax. Finally, note that if all attempts to increase traction fail, the remedy is to rewax the skis under the foot with a relatively thick application of a softer wax, presumably because the thicker, softer wax more readily accepts embedding by the snow grains [10].

Recall from Fig. 8.8 above that the hardness of ice decreases markedly at temperatures ranging from −5° to the melting point of ice, 0 °C, and that the hardness of ice increases by a factor of about 2 as it cools from −5 to −15 °C. Given these physical characteristics of the hardness of ice, a soft, klister wax used for traction at −2 °C or above must be very soft indeed if the ice is to embed in it and produce good grab. As the temperature falls from −2 to −5 °C and below, most climbing or kicker waxes that a skier may use should become somewhat harder and still remain serviceable. However, any wax that is too soft or applied too thickly will allow the harder snow grains that exist at colder temperatures to embed more deeply, until the snow grains fail to release from the wax and the resulting buildup of snow and ice negates the glide phase of the grab-and-glide cycle.

Finally, consider also that the ability of snow grains to embed in a ski wax should also depend on the hardness of the base material of the ski. Harder base materials probably require thicker, more pliable layers of wax. This conjecture is born out by the experience of skiers who use old-fashioned, wood nordic skis. It is uniformly observed that harder waxes will serve for traction on wood skis than will serve on modern skis with harder, polyethylene bases. Nordic track ski racers have also found that a layer of relatively soft wax overlain with a layer of harder glide wax will serve for the combination of kick traction and good glide they require.

FRICTION AS ART

In a sense, much of skiing involves a complex, intriguing engagement and disengagement with the many processes that control friction. When the goal is solely to promote good glide, at low speeds the friction process primarily involves solid-to-solid contact of snow asperites with the hard, waxed running surface of the ski. As the glide speeds increase, other processes join in, most notably a lubricating film of meltwater that operates in parallel with the abrasive, solid-to-solid friction process. For nordic skiing, proper waxing that will produce an effective grab-and-glide cycle requires an artful balancing of wax hardness (and softness), the placement of the wax on the ski (for more grab, on the running surface under the boot where the normal force is strongest), and the thickness of the wax layer as the wax interacts with the hardness (and softness) of the snow grains, which is controlled by the temperature. The variables are many and the interactions of the variables are complex.

We have noted many of the physical properties and processes involved in the interplay of glide and grab, but finally all skiers probably can, at best, use this understanding only to augment and, perhaps, refine the knowledge they gain from their own experience. For skiers, working with the many processes of friction, which means applying the proper wax at the proper time in the proper place and in the proper amount, remains largely an art.

REFERENCES

1. S. C. Colbeck, in "A Review of the Processes that Control Snow Friction," CRREL Monograph 92-2 (U.S. Army Corps of Engineers, Cold Regions Research and Engineering Laboratory, Hanover, NH, 1992), offers an extensive summary and review of the scientific work on this subject done to the date of his article, citing 116 references; see also his subsequent bibliography of sources on the topic, "Bibliography on Snow and Ice Friction," CRREL Special Report No. 93-6 (U.S. Army Corps of Engineers, Cold Regions Research and Engineering Laboratory, Hanover, NH, 1993). R. Perla and B. Glenne, in their essay, "Skiing," in *Handbook of Snow*, edited by D. M. Gray and D. H. Male (Pergamon, Toronto, 1981), offer excellent anecdotal advice regarding waxes and running surface configurations.
2. This discussion derives from the work of B. Glenne, "Sliding Friction and Boundary Lubrication of Snow," Trans. ASME J. Tribol. **109**(4), 616 (1987).
3. The values given are from the work of S. C. Colbeck, cited in Ref. 1 (1992).
4. W. Ambach and B. Mayr report these observations in "Ski Gliding and Water Film," Cold Regions Sci. Technol. **5**, 59 (1981).
5. G. C. Warren, S. C. Colbeck, and F. E. Kennedy report these measurements in "Thermal Response of Downhill Skis," Report No. 89-23 (U.S. Army Corps of Engineers, Cold Regions Research and Engineering Laboratory, Hanover, NH, 1989).
6. Figure 8.4 is from S. C. Colbeck's [Ref. 1 (1992)] work on snow friction. Colbeck used a fiber block to simulate the bottom of a ski.
7. For example, L. Torgerśen's book, *Good Glide: The Science of Ski Waxing* (Human Kinetics, Champaign, IL, 1983), offers useful discussions of the techniques for waxing skis to best address a variety of conditions and performance requirements, but very little of the underlying science is considered. The "science" in the title refers mostly to practices that have been developed from extensive experience.
8. The literature available from commercial wax manufacturers focuses almost entirely on the "how to" and "when to" application aspects of their products. In the discussion that follows, we are indebted to one of the few scientific discussions of the chemistry of ski waxes, a monograph by G. M. Street and P. Tsui, "Composition of Glide Waxes

Used in Cross Country Skiing,'' (Pennsylvania State University, Biomechanics Laboratory, 1987). Glide waxes, whether they are used for alpine or for cross-country skis, are essentially the same.

9. This finding is reported in the monograph by G. M Street and P. Tsui (1987) cited in Ref. 8.

10. M. Shimbo, in "Friction of Snow on Ski Soles, Unwaxed and Waxed," in *Scientific Study of Skiing in Japan*, edited by K. Kinosita (Hitachi, Tokyo, 1971), reports coefficients of sliding and static friction as a function of wax thickness, which supports the idea that snow grains embedding in the wax base preparation is the essential phenomenon that produces traction in the grab-and-glide cycle. Interested readers should also see the discussions of ice hardness in F. P. Bowden and D. Tabor, *The Friction and Lubrication of Solids* (Clarendon, Oxford, 1964), Part II.

CHAPTER 9

EPILOGUE
Physics, Skiing, and the Future

In a sense, the speed and intensity of today's research and development efforts regarding the physical properties of ski manufacturing materials and ski design have brought the future much closer to us. Skiers who use the latest equipment built for the current season are, in relation to the many more skiers on the hill whose equipment is even just a few years old, the skiers of the future. They use skis made of composite materials designed specifically to address the flexural, torsional, and vibrational demands of skiing, and those skis are designed into special shapes and contours that promote enhanced glide and ever tighter turns made ever faster and carved with ever greater control.

It seems that the more we understand about the physical forces associated with the mechanics of carving turns, the more skiing and snowboarding have evolved so that they share some important characteristics of both equipment and technique. Skis have become shorter and lighter, greatly reducing their swing weights. They have narrower waists and wider shovels and tails, giving them greatly increased sidecuts. Skiers using these futuristic, parabolic skis carve tight turns in the snow, angulating their bodies in attitudes not far removed from the snowboarders' trademark angulation, so acute that they can push off against the snow with their gloves as they link their turns, changing edges in an instant. Skiers and ski instructors have had to scramble to adjust to the changes in skiing technique brought to our sport by the ever-increasing variety of new equipment.

The recent past can tell us something about what to expect from the near future regarding changes in equipment and technique. Over the last decade the introduction of the skating technique has revolutionized nordic track racing equipment and technique, relegating those who use the traditional, diagonal stride racing skis and technique to their own, increasingly nostalgic, places in the schedules of racing events and in the record books. It was only after-the-fact analysis and understanding of the underlying physics of propulsion and the biomechanical physics of the body's capacity for work that made the benefits of the new, nordic ski-skating technique obvious. We can only wonder why someone did not think of using the skating technique sooner, and we can only speculate that someone might have done just that had he or she analyzed the physics of nordic skiing more closely more years ago. Other such breakthroughs are not only possible, but likely, as researchers consider the human body as a machine for doing work and add a growing understanding of the kinesiology of the body (and its limits) to their understanding of the physics of equipment and technique.

Perhaps the most important development along these lines for the recreational skier is increased understanding of the physics of injury in all sports, including skiing. The development of modern boots and safety bindings has been driven, until fairly recently, by the experiences of skiers—both professional racers and recreational skiers—on the hill. Through this process of experience and experiment, modern boots with their molded plastic shells have been designed to distribute the loads that act on the skier's lower body higher up on the calf and to the knee, with the result that the incidence of ankle and lower leg injury (once among the most common of ski injuries) has been greatly reduced.

Unfortunately, the knee has always been the part of the skier's body most at risk for injury, and with modern boots and bindings tending to carry the stresses on the lower body to the knee, the incidence of sprained knees and torn ligaments is higher now than it has been in the past [1]. The more we know about the physics of knee injuries, the more we see that perhaps the most effective way to approach this problem is through prevention. The knee lends itself well to instrumentation, and several studies have been made that measure the forces and torques generated by skiing (and by falling) that act on the binding, the boot, the lower leg or calf (the tibia and fibula bones), the knee, and the upper leg or thigh (the femur) [2].

First we should look at the release mechanisms of modern safety bindings and their recommended adjustments. Researchers have found that the force required to release the toe binding correlates well with the lateral force applied to a ski on a test bench. Applying lateral force to the ski is a good measure of the axial torque—the twisting, rotational force—applied to the lower leg (principally the larger bone in the calf, the tibia) in a fall when the ski's edge does not release and the upper body twists. But when tests were done in the field, the moment and force required to make the toe binding release were 33% larger than the bench-tested recommended settings. On the ski slope it is possible that flexural bending of the ski may load the boot in the binding and inhibit the release mechanism. Similarly, during field tests of the forward–backward moment at the boot required to release the heel binding, the force required exceeded the bench-tested recommended setting by 100%. This result indicates that the heel force between the boot and the binding as set is not a good indicator of the actual moment the boot may exert on the tibia of the lower leg. In most of the field tests conducted in this research, the recorded moments required at the boot to release the binding in a variety of field situations exceeded the static release settings in 50% of the cases, and the moments recorded in the field ranged as high as 2.4 times the moment that, according to the recommended settings, should have caused the binding to release. Given these findings, equipment manufacturers need to implement standards for release settings based on field studies, not bench studies. Some design improvements should also be made that take into account the loading of the boot against the ski and the vertical force at the toe and how those forces may inhibit the release mechanism of a safety binding.

Finally, all skiers, especially recreational alpine skiers, should probably heed the warnings that are made explicit by biomechanical studies of the physics of knee injuries. Certainly the recreational skier can ski so as to avoid injury and still have an enjoyable, exhilarating experience [3]. Simply having the common sense to know realistically what one's ability is and then to ski within one's ability is the greatest part of skiing so as to avoid injury. Knowing something about the physics of ski injuries—something about the forces that act in a fall—might help skiers be more realistic in assessing their skiing comfort zones. Because knee injury, in particular tearing or straining the anterior cruciate ligament, is one of the more common ski injuries, let us look more closely at the physical processes at work.

The knee joint articulates the bones of the leg: the femur above and the tibia and fibula below. The knee joint is held in place by several ligaments and two pads of cartilage—the medial meniscus on the inside and the lateral meniscus on the outside—that cushion the coming together of the two larger bones, the femur and the tibia. The two cruciate ligaments (so named because they cross each other) are situated between the two menisci but not in the knee joint capsule itself. The anterior cruciate ligament (ACL) connects from the front center of the tibia upward to the back outside of the femur. The posterior cruciate ligament (PCL) connects from the back center of the tibia upward to the front center of the femur. Most household medical references include diagrams of the knee joint and its associated ligaments. The ACL keeps the tibia from sliding forward or from moving too far to the inside. Because the ACL is connected diagonally across the knee joint from the front and center of the tibia to the back and outside of the femur, it is subject to stretching when the ski and the lower leg twists to point inward.

Let us consider the forces that occur in a specific fall [4]. Figure 9.1 shows the bones of the leg and the forces generated when a skier using modern safety bindings and wearing boots that reach to the mid- to upper calf tries to recover from a backward fall. Only sideways motion will cause the toe binding to release, and only upward loading will release the heel binding. The skier's knee flexes in the manner illustrated as the skier tries to recover his balance and avoid falling. The skier's center of mass (CM) is to the rear, so the skier's weight W is shown acting downward, creating the opposing reaction force S_h directed upward at the heel and its associated force S_t, which is directed downward at the toe. The net upward force on the foot, $S_h - S_t$, is equal to W. Look at the forces acting on the tibia and the femur. At the knee, the quadriceps muscle from the upper thigh exerts a force Q that extends over the kneecap to the front of the tibia, while the hamstring muscle of the lower thigh exerts the force H that pulls the tibia backward, as shown. The contact pressure of the femur on the tibia at the

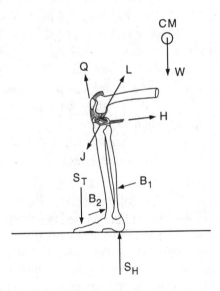

FIGURE 9.1. The bones of the lower leg, the knee joint with the crossed anterior and posterior cruciate ligaments, and the femur are shown as they are stressed by a backward fall from which the skier attempts to recover his balance. (Adapted from Figueras et al., 1987. Copyright ASTM. Reprinted with permission.)

knee joint is shown as the force J, directed downward at an angle, and the tension force induced in the ACL is designated as L. Remember, modern ski equipment makes the boot, the ski, and the lower and upper leg form an essentially rigid unit. Given the directions and the likely magnitudes of these forces (depending on the skier's weight and velocity), in almost any fall of this sort, something has got to give. The quadriceps and the hamstring forces must be quite large in order to overcome the moment caused by the weight W and thus maintain the skier's balance. If the hamstring force H lags somewhat below the force imposed by the quadriceps force Q, the tibia will move forward, which places an excessive load on the ACL, leading to its failure.

When we ski with some velocity, the physical limitations of our body's reaction times often will simply not allow us to monitor our muscles in time to make appropriate corrections when we are thrown off balance—even if we assume that we have the degree of musculature required to deal with the forces that a fall such as the one we have discussed might generate. Knowing how an injury may occur and the physics of the forces that may be generated by a particular planned or unplanned maneuver should encourage recreational skiers to ski defensively, to ski so as to avoid injury.

In the case of the backward fall discussed above, the skier might have decided simply to bail out by rotating his body and knees sideways and falling into the hill, in so doing absorbing the psychological indignity of a spill rather than risking the very real threat of serious physical injury. In a less drastic but still serious context, we see the concept of skiing so as to avoid injury in the ski instructor's common practice of teaching beginning skiers as soon as possible to use a skater's stop, not a snowplow, when they are skiing with any speed at all and they want to put on the brakes. The ACL is commonly injured when the knee and the leg are twisted inward. Such an inward twist can easily happen when the skis are in a modified wedge position, as they are when a skier tries to stop by using the snowplow maneuver.

Finally, we firmly believe that everyone associated with skiing— recreational and professional skiers; alpine, nordic, and adventure skiers; ski-resort owners and lift operators; ski-equipment manufacturers and retailers—might benefit from paying more attention to the physics of skiing. The understanding that comes from knowing something about the how and why of the physical forces that attend any aspect of skiing can only add to our enjoyment and to our successes as we pursue this great sport into a future that promises an increasingly brisk pace of continuous change and positive transformation.

REFERENCES

1. Interested readers should see L. R. Young and S. M. Lee, "Alpine Injury Pattern at Waterville Valley, 1989 Update," in *Ski Trauma and Safety: Eighth International Symposium*, ASTM STP 1104 (American Society for Testing and Materials, Philadelphia, 1991), pp. 125–132, which offers tables that show the distribution of injuries for beginner, intermediate, and expert skiers. The incidence of knee injury is the highest for all three classes.
2. Interested readers will find much detailed information on this topic in J. M. Figueras, F. Escalas, A.Vidal, R. Morgenstern, J. M. Bulo, J. A. Merino, and J. M. Espadaler-Gamisans, "The Anterior Cruciate Ligament Injury in Skiers," in *Ski Trauma and Safety: Sixth International Symposium*, ASTM STP 938 (American Society for Testing and Materials, Philadelphia, 1987), pp. 55–60; see also C. Y. Kuo, J. K. Louie, and C. D. Mote, "Control of Torsion and Bending of the Lower Extremity During Skiing," in *Ski Trauma and Safety: Fifth International Symposium*, ASTM STP 860 (American Society for Testing and Materials, Philadelphia, 1985), pp. 91–109. The discussion that follows derives from the work presented in these sources.
3. See the work of C. F. Ettlinger and his colleagues at Vermont Safety Research, PO Box 85, Underhill Center, Vermont 05490 for specific information on "skiing so as to avoid injury." See also C. F. Ettlinger, R. J. Johnson, and J. E. Shealey, "A Method to Help Reduce the Risk of Serious Knee Sprains Incurred in Alpine Skiing," Am. J. of Sports Med. 23(5), pp. 531–537 (1995).
4. This example is drawn from the work of J. M. Figueras, *et al.* cited in Ref. 2.

THERMODYNAMICS OF PHASE CHANGES

Reviewing the thermodynamics of the phase changes that occur in water, especially over curved phase boundaries, will help us appreciate the phenomena associated with the formation of snow in the atmosphere and the changes that occur in a snowpack [1].

Starting with a model of a perfect gas in a container, the relation between the pressure P, the volume V, and the absolute temperature T, given in kelvins, is the following:

$$PV = nRT. \qquad \text{(T1.1)}$$

R is the gas constant and n is the number of moles of gas contained in the vessel. If m is the mass of gas, $n = m/M$, where M is the molecular weight and ρ, the gas density, equals m/V, so we may write the gas law as

$$PV = \frac{mRT}{M} \rightarrow P = \frac{\rho TR}{M}. \qquad \text{(T1.2)}$$

The thermodynamic principles are contained in the following laws. Heat is a form of energy. For a system contained in a cylinder in contact with a source of heat (the example given in Fig. 2.16, p. 37) when heat Q is transferred, work W may be done, and the internal energy U of the system may change. The first law of thermodynamics states

$$\Delta Q = dU + dW. \qquad \text{(T1.3)}$$

The internal energy is a function of T, V, and n for a perfect gas through the equation of state. The work done by the pressure of the gas on the piston is $dW = PdV$ in the expansion process. When the gas is compressed, $dW = -PdV$. The first law of thermodynamics states that the heat added to the system changes the internal energy and contributes to the work done by the system. The process may be isothermal $(dT=0)$ isobaric $(dP=0)$, or adiabatic, i.e., thermally isolated $(\Delta Q=0)$.

The second law of thermodynamics puts a quantitative restriction on the conversion of heat energy to work. A new thermodynamic function, entropy S, is defined as $dS = \Delta Q/T$. The second law states that

$$S_B - S_A \geq \int_A^B \frac{\Delta Q}{T}. \tag{T1.4}$$

The equal sign is valid only for reversible processes or for conditions of thermodynamic equilibrium. The entropy difference between states A and B is independent of the path taken, as long as all processes are reversible. For irreversible processes, the entropy difference is always greater than the integral expression.

Let us define the thermodynamic state functions. From the first law, note that the internal energy U, which is the sum of all kinetic and potential energies of the molecules in the system, depends on the variables S and V as follows:

$$\Delta Q = dU + PdV = TdS,$$

$$dU = TdS - PdV, \tag{T1.5}$$

so

$$U = U(S,V).$$

The internal energy is a function only of the entropy S and the volume V of the system. By adding and subtracting VdP to dU, we get

$$dU = TdS - d(PV) + VdP,$$

$$d(U+PV) = TdS + VdP = dH, \tag{T1.6}$$

so

$$H = U + PV = H(S,P).$$

The state function $H(S,P)$ is called the heat function or enthalpy.

Now let us look at the other function associated with this process, the Gibbs free energy function. Add and subtract SdT to dH:

$$dH = TdS + SdT + VdP - SdT = d(ST) + VdP - SdT,$$

so

$$d(H - ST) = VdP - SdT = dG, \tag{T1.7}$$

$$G(P,T) = H - ST = U + PV - ST.$$

Thus T, V, P, and S are thermodynamic variables; U, H, and G are state functions defined by the variables. T and P do not depend on mass or composition; all of the other quantities do. For a single-component system

$$G = G(T,P,N),$$

$$H = H(S,P,N), \tag{T1.8}$$

$$U = U(S,V,N),$$

where N is the number of moles in the system. The following principles govern the trend toward equilibrium:

(a) Every isolated system tends toward equilibrium.

(b) Internal energy U, like potential energy, tends to decrease in an isolated system.

(c) From the second law, entropy, or S, tends to increase toward a maximum; thus at equilibrium $G = U - TS + PV$ becomes a minimum value.

Now write $dG = -SdT + VdP + \mu dN$; μ is the chemical potential, a measure of the energy stored in chemical bonds. Apply this relation to a system of water at one of the phase boundaries where, for example, liquid and vapor coexist at a pressure P and temperature T. Condense some vapor, dN_v, to liquid, dN_l; dT and dP are unchanged, and since G is a minimum, $dG = 0$. Thus $\mu_l = \mu_v$. The chemical potential of the vapor and liquid are the same at the phase boundary because the water molecule has not been altered. Let us apply these results to water when the liquid and vapor states are in equilibrium.

Refer to Fig. 2.18, p. 41, the phase diagram for water near the triple point. Consider the transition, moving from a to b on the liquid (l) side of the line and then on the vapor (v) side. The expressions for the two phases are written as follows:

$$G_l = U_l + PV_l - TS_l,$$

also

$$G_v = U_v + PV_v - TS_v. \tag{T1.9}$$

Now, make an imaginary excursion along the phase boundary line from a to b for the liquid and vapor, then

$$dG_l = V_l dP - S_l dT,$$

also   (T1.10)

$$dG_v = V_v dP - S_v dT.$$

Since the phases are in equilibrium, $G_l = G_v$, always; thus

$$V_l dP - S_l dT = V_v dP - S_v dT,$$
                                                                        (T1.11)
$$\frac{dP}{dT} = \frac{S_v - S_l}{V_v - V_l}.$$

Suppose that heat is supplied to the system when the liquid and vapor are in equilibrium; P and T do not change, but the heat added causes the liquid to evaporate to vapor. The heat added to the system, ΔQ, is equal to the heat of vaporization, L, which is equal to $T(S_v - S_l)$. Thus the heat of vaporization, L, results in the entropy change when the liquid changes to vapor. L is usually given for one mole, so the volumes are molal volumes also.

The expression for the basic Clausius Clapeyron equation is given above in the form represented in Eq. (T1.11). Because $S_v - S_l = L_{lv}/T$,

$$\frac{dP}{dT} = \frac{L_{lv}}{T(V_v - V_l)}.$$ (T1.12)

The volume of a mole of vapor is much larger than the volume of a mole of liquid: $V_v \gg V_l$. Now, using $PV_v = RT$, Eq. (T1.12) becomes

$$\frac{dP}{dT} = \frac{LP}{RT^2} \rightarrow \frac{dP}{P} = \frac{L dT}{RT^2}.$$ (T1.13)

Integrating expression (T1.13) yields the value of P at any point on the phase curve relative to some reference point. Since $V_v - V_l > 0$, the slope is positive.

There is a similar relation for the liquid to solid phase curve:

$$\frac{dP}{dT} = \frac{F}{T(V_l - V_s)}.$$ (T1.14)

F is the molal heat of fusion, or the heat needed to change the solid to the liquid. For ice, V_s is larger than V_l, so the slope is negative. Therefore it follows that ice when put under pressure must melt. The change in the melting point is 0.0075 °C for one atmosphere of pressure. This accounts in part for the water film that lubricates an ice skater's blade as it bears down on the ice. F is 80 cal/g for ice, while L is 596 and 540 cal/g at 0° and 100 °C respectively for liquid water.

For sublimation of the solid to vapor at 0 °C, the heat of vaporization is 80+596 or 676 cal/g, so the slope of the solid to vapor curve is greater than the supercooled liquid to vapor curve below the triple point. Thus the vapor pressure over the solid is always less than the vapor pressure over the supercooled liquid, so snow grows at the expense of fog in the atmosphere.

CLAUSIUS–CLAPEYRON RELATION FOR CURVED SURFACES

The basic Clausius–Clapeyron relation applied to the phase transitions above is valid for phase boundaries that occur on plane surfaces. The relation must be modified when we consider water droplets and ice crystals because all liquids and solids have surface energies that change the pressure values in the different phases. Figure T1.1 shows a spherical, liquid droplet of water under additional pressure produced by the surface tension σ.

The total force holding the droplet together around a circumference must be equal to the pressure force that would blow it apart; thus

$$2\pi r\sigma = \pi r^2 P_i \rightarrow P_i = \frac{2\sigma}{r}. \tag{T1.15}$$

The additional energy of the water inside the droplet is $P_i V$, so the heat of vaporization needed is $L_i = L - P_i V$ for one mole. Modify expression (T1.13) accordingly, and we get

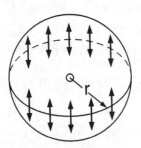

FIGURE T1.1. Spherical droplet of liquid water with a radius r. The arrows represent surface-tension forces that compress the droplet around a circumference.

$$\frac{dP}{P} = \frac{L_i}{R}\frac{dT}{T^2} = \frac{L - P_i V}{R}\frac{dT}{T^2},$$

(T1.16)

$$\frac{dP}{P} = \frac{L'}{R'}\frac{dT}{T^2} - \frac{P_i}{R'\rho_i}\frac{dT}{T^2}.$$

If L_i and R are divided by M, the molecular weight, L' and R' are the corresponding values per gram, $V/M = 1/\rho_i$, and the second expression in (T1.16) results. The differential equation is then integrated from initial T_0 and P_0 to T and P, giving the relation

$$\ln\frac{P}{P_0} = \frac{L'}{R'}\left(\frac{1}{T_0} - \frac{1}{T}\right) + \frac{P_i}{R'\rho_i}\frac{1}{T}.$$

(T1.17)

When we write (T1.17) as an exponential, it becomes the following:

$$P = P_0 e^{(L'/R')(1/T_0 - 1/T)} e^{P_i/R'\rho_i T}.$$

(T1.18)

The first exponential term is the vapor pressure over a flat surface at some temperature T. Label it $P(T)$, and then write the expression as

$$P = P(T)e^{P_i/R'\rho_i T}$$

(T1.19)

When P_i is replaced by the value in terms of the radius r of the droplet, the final result becomes

$$P = P(T)e^{2\sigma/r\rho_i R'T}.$$

(T1.20)

As the radius of the droplet decreases, the effective pressure inside the droplet increases. The value of the radius is positive if the surface is convex. For a concave surface, the value of r is negative, so the vapor pressure is lower than it is over a plane surface. Thus water evaporates from convex surfaces and condenses at concave surfaces. A precisely similar relation, with a different value of σ, applies for sublimation from the solid ice to the vapor. Again, water migrates from the sharp points on the ice grains to the flat areas or to the concave areas created at the contact points between two ice grains.

A similar effect occurs at the interface between the solid and liquid phases. The result is expressed as a depression of the melting point ΔT_m from the temperature T_0 as the result of the convex curvature of the solid and a rise in the melting point at the concave regions. The relation is expressed as follows:

$$\Delta T_m = -\frac{2T_0\sigma_{sl}}{L_{sl}\rho_s r_s}. \tag{T1.21}$$

Ice grains with small radii have lower melting points than do the larger grains, and thus they will melt, because heat for melting is conducted by the liquid from the large grains as the liquid freezes. Also, the ice grains melt at the concavities and refreeze at the convexities, so any bonding of grains is destroyed. The ice grains become spherical, and under these conditions the snowpack becomes completely unstable.

REFERENCE

1. See S. C. Colbeck, "Introduction to the Basic Thermodynamics of Cold Capillary Systems," CRREL Special Report No. 81-6 (U.S. Army Corps of Engineers, Cold Regions Research and Engineering Laboratory, Hanover, NH, 1981), who treats the problems discussed here.

SKI LOADING AND FLEXURE ON A GROOMED SNOW SURFACE

The American Society of Testing and Materials (ASTM) defines the bending stiffness of a ski by its deflection under a given load placed on the ski at its midpoint when the ski is supported at its tip and tail contact points. For a casual comparison of ski flexure values, the values obtained in this manner may be adequate. To determine a ski's loading distribution and flexure in a real snow bed, we need to consider the distribution of the ski's bending stiffness, which may be described by the relationships associated with the mechanics of a nonuniform beam in flexure [1]. A similar set of relations applies for analyzing the torsional stiffness of a ski; however, in this discussion we consider coupled bending and torsional deflections only qualitatively. We address the specific problem of predicting the edge loading on a ski when it bends to contact the packed and groomed surface of an alpine ski slope.

Refer to Figs. T2.1(a) and T2.1(b), which illustrate two views of sections of a flexed beam—or ski. First consider the beam under bending stress only. Equal and opposite moments are applied at the ends of the beam to give a curvature of $1/R = d^2y/dx^2$. The moment is then related to the curvature by a bending stiffness coefficient B, and it follows that

$$M = \frac{B}{R} = B\,\frac{d^2y}{dx^2}$$

$$(T2.1)$$

$$B = \int E(y)y^2\,dA.$$

199

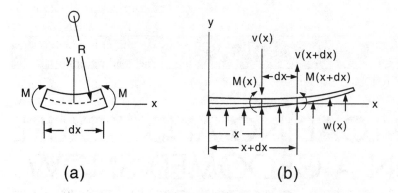

FIGURE T2.1. Two views of a flexed beam, or ski. View (a) shows the moment M that produces the curvature of radius R, depending on the beam stiffness B. View (b) represents the force and moment diagram to determine the relations between the shear force $V(x)$ and the moment distribution $M(x)$ in terms of the distributed load $w(x)$.

The value of B given above is the expression for a composite beam, in which dA is the element of area at the distance y from the neutral axis, and $E(y)$ is the appropriate rigidity modulus for the material from which the beam is constructed. If the beam is of uniform composition, then $B=EI$, where $I=\int y^2 dA$ is the beam moment. The bending stiffness coefficient B is always calculated with respect to the neutral axis of the section, that is, the position in the beam where the stress field changes from tension to compression. For the purposes of our discussion, we measure the value of B as a function of position along the ski, because the construction and composition of actual skis is quite varied. For the results of such measures, see Fig. 3.4, p. 58, where the bending stiffness B of four skis is plotted for every 10 cm from the center points of the skis.

Now consider the ski element shown in Fig. T2.1(b), which is supported by a groomed snow slope. By consideration of the forces and torques acting on each section of the ski, the shearing force V across the ski section follows the relation $dV/dx = -w$, and the bending moment becomes $dM/dx = -V$, where w is the distributed load per unit length along the ski. These relations then lead to the differential equation for the beam deflection given by

$$\frac{d^2M}{dx^2} = w \rightarrow \frac{d^2}{dx^2}\left(B\,\frac{d^2y}{dx^2}\right) = w, \qquad \text{(T2.2)}$$

where B depends on x, the linear position on the beam, and w depends on both x and the beam deflection y. Note that $w = W(x)p(y)$, where $W(x)$ is the width and $p(y)$ is the pressure of the snow on the ski, which varies according to the deformation of the snow due to the loading applied.

We may treat a packed and groomed alpine ski slope as an elastic foundation, so that $p = Ky$; that is, the pressure p is proportional to the snow deformation, and K is a constant of proportionality. Snow deformation on a packed slope, however, is usually negligible compared to changes in the contour of the snow surface, so the actual $y(x)$ function for the ski on a groomed snow slope is specified by the shape of the surface with which the ski makes contact and does not involve any surface deformation. Given these assumptions, the differential equation (T2.2) may be used to evaluate w, the loading pressure along the ski.

Assume that the loading pressure along the ski is sufficient to force contact with the snow over the entire length of the ski. Note that w is the snow reaction loading force along the ski edge, and that $M = B/R$. Assume that the radius of curvature, R, is constant. Thus w is proportional to the second derivative, or the curvature of the B function along the ski. From the examples for the bending stiffness of skis graphed for the B function in Chap. 3, Fig. 3.4, p. 58, observe that the curvatures illustrated are positive at the tip and tail regions of the skis and negative in the center regions. Thus an upward pressure must be applied at the tip and tail of the ski and a downward pressure must be applied in the center region to achieve this distribution. Such a load distribution is shown in Fig. 3.7, p. 64, which graphs the load distribution for a Head ski.

The ski boot applies an external load to the ski that keeps the ski in static equilibrium. Boot loading is usually more than the amount required just to flex the ski. Part of the boot force supplies the downward center load; the remainder then generates an additional upward snow-reaction force under the boot. In every case, the sum of the snow-reaction forces on the ski must be just equal to the boot-load force. The placement of the boot load must be such that there is no torque acting on the ski.

Note also that if the ski has some initial camber, the effective value of R is reduced. The resulting curvature of the ski under a load is the sum of the camber curvature and the contour curvature: the tip and tail loadings increase and the loading under the boot is reduced. Such a load distribution enhances the ski's stability at high speed, which suggests that being able to adjust the camber of a ski, as well as its center stiffness, would be desirable.

REFERENCE

1. For a useful discussion of the mechanics of beams in flexure and torsion, see S. Timoshenko and J. N. Goodier, *Theory of Elasticity*, 2nd ed. (McGraw-Hill, New York, 1951), pp. 70–73.

THE LOADS ON A RUNNING SKI

To analyze the forces acting on a ski during a traverse, refer to Figs. 4.5(a) and 4.5(b) in Chap. 4, p. 90, which depicts a skier running down an incline at an angle β from a horizontal line on the slope. Figure 4.5(b) illustrates the force diagram for the case illustrated in Fig. 4.5(a). The skier's weight **W** may be resolved into force components normal to the slope, $\mathbf{F}_N$, and down the fall line, $\mathbf{F}_S$, as follows:

$$\mathbf{W} = \mathbf{F}_N + \mathbf{F}_S,$$

$$F_N = W \cos \alpha, \quad F_S = W \sin \alpha. \tag{T3.1}$$

The force $\mathbf{F}_S$ is further resolved into components that apply along the direction of motion, $\mathbf{F}_p$, and perpendicular to the ski, $\mathbf{F}_{\text{lat}}$:

$$\mathbf{F}_S = \mathbf{F}_p + \mathbf{F}_{\text{lat}},$$

$$F_p = W \sin \alpha \sin \beta, \quad F_{\text{lat}} = W \sin \alpha \cos \beta. \tag{T3.2}$$

From Newton's equations of motion note that an inertial force, $\mathbf{F}_I = -M\mathbf{a}$, with a direction opposite to the acceleration must be added to the force diagram to represent any acceleration. This force, $\mathbf{F}_I$, is shown acting at the CM, and it is just equal and opposite to the force $\mathbf{F}_p$ that accelerates the skier along his track. The force $\mathbf{F}_p$ also acts at the center of mass, but it is shown in the diagram at the ski to indicate that it is along the direction of the ski. The force $\mathbf{F}_{\text{lat}}$ is a sideways force in the slope that causes the skis to skid to the side and out of the direction of the track. Without sideways motion, the skier's biomechanical balance system will tilt the skier so that the sum of the forces $\mathbf{F}_N$ and $\mathbf{F}_{\text{lat}}$ is directly through the feet and hence normal to the ski. This direction then defines the orientation of the ski plane. Another force $\mathbf{R}$ represents the upward snow reaction needed to balance the force $\mathbf{F}_{\text{load}} = \mathbf{F}_N + \mathbf{F}_{\text{lat}}$. If the skier's track were directly down the fall line, then the direction of $\mathbf{F}_N$ would be the true direction for balance.

From Fig. 4.5(b), p. 90, the reader may see that $\mathbf{F}_{\text{lat}}$ is perpendicular to $\mathbf{F}_N$, so

$$\tan \phi = \frac{W \sin \alpha \cos \beta}{W \cos \alpha} = \tan \alpha \cos \beta,$$

$$F_{\text{load}} = W(\cos^2 \alpha + \sin^2 \alpha \cos^2 \beta)^{1/2}.$$

(T3.3)

The force F_{load} determines the edge loading of the ski and hence the cutting action of the edge on the surface of the snow.

When the direction of $\mathbf{F}_{\text{load}}$ is along the leg and passes through the ski [as it does, for example, in Figs. 4.11(a) and 4.11(b), p. 101], the angle Φ between the plane of the ski and the slope is defined by the angle ϕ between the normal to the slope $\mathbf{F}_N$ and $\mathbf{F}_{\text{load}}$.

GEOMETRY OF THE EDGED AND FLEXED SKI

Refer to Fig. T4.1, which shows the outline of a conventional ski as seen from the top view. The contact length between the shovel and tail is labeled C. The tail width is T, the waist width is W, and the shovel width is S. We assume that the waist of the ski, where W is measured, is near the midpoint of the contact line. The sidecut, SC, is given by the relation $SC=(S+T-2W)/4$. From the geometry illustrated in Fig. T4.1, we can write $C\approx\delta R_{SC}$, and then expand the cosine function to use the first two terms only, as follows:

$$SC=R_{SC}\left[1-\cos\left(\frac{\delta}{2}\right)\right]\approx R_{SC}\left[1-1+\frac{1}{2}\left(\frac{\delta}{2}\right)^{2}\right], \tag{T4.1}$$

$$SC=\frac{R_{SC}\delta^{2}}{8}=\frac{C^{2}}{8R_{SC}},$$

thus $\hspace{10cm}$ (T4.2)

$$R_{SC}=\frac{C^{2}}{8SC}.$$

From the geometry, we have expressed the approximate relations of the sidecut radius R_{SC} in terms of the sidecut SC.

Now, let us consider the deflection of a ski with its edge in contact with a slope. Refer to Fig. 3.6(b), p. 63, in which we look along the shovel to tail contact line of a ski edge into a slope. A circle that passes through the edge points defined by the shovel, the waist, and the tail of the ski has a radius that will define the contact line of the ski where it presses into the slope and hence the carve radius R_{con}. Before deflection, the waist section

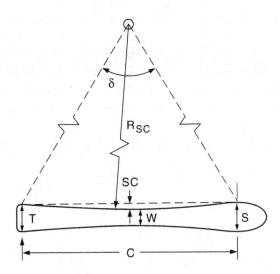

FIGURE T4.1. Geometry of the ski showing the relation between the sidecut and the sidecut radius.

of the ski is above the contact line. From the geometry, note that the radius of the arc of the line of the ski edge when it is deflected into contact with the slope is given by

$$R_{con} = \frac{C^2 \cos \Phi}{8SC} = R_{SC} \cos \Phi. \tag{T4.3}$$

This expression gives us the approximate radius of the carving edge of the ski on the surface shown because neither the edge profile nor the contact line will make an exact arc of a circle.

The actual center displacement of the ski is SC tan $\Phi + c$, where c is the natural camber height of the ski. This deflection determines the distributed loading from the shovel to the tail of the ski needed for the ski to make contact with the slope. The remainder of the weight is distributed under the boot, depending on the relative stiffness of the ski and hardness of the snow bed. Because $d^2y/dx^2 = 1/R_{flex}$, where y is the deflection as a function of the position x along the ski, and R_{flex} is the radius of flexure, we can calculate the loading directly from R_{flex} by the development given in Technote 2, p. 199:

$$R_{flex} = \frac{C^2}{8(SC \tan \Phi + C)}. \tag{T4.4}$$

If the skier uses no angulation, which in this context would be lateral rotation of the skis away from a flat-footed stance, the angle Φ is the same as ϕ. Remember that the angle ϕ is the angle between the perpendicular reaction force on the ski ($\mathbf{F}_{load}$) and the normal of the slope. The value of R_{con} determines the radius of a carved turn, and R_{flex} determines the distribution of loading on the edge of the ski during that turn.

THE DYNAMICS
OF CARVING A TURN

The perfectly carved turn requires that the skier remain in a balanced position. The forces acting on the skier are the force of gravity, that is, the skier's weight at the center of mass, aerodynamic friction from the air passing over the skier's body, and an inertial force parallel but opposite in direction to the skier's acceleration. The snow bed provides reaction forces: a force normal to the plane of the slope; a force in the slope plane with a direction perpendicular to the ski edge, which acts as a chisel; and, finally, a force parallel to the ski that results from the friction produced from the ski's sliding over the snow surface.

To be in perfect balance, the sum of all these forces acting on the skier, including the inertial forces, must be zero; the torque about the center of mass produced by all of these forces must also be zero. The skier's biomechanical feedback seeks body balance, and it causes the skier to set his feet at a position under his center of mass so that the conditions above regarding the balancing of forces are satisfied. In other words, the skier uses body control where it is needed as he uses different maneuvers to negotiate the different demands of the snow and the slope.

For this discussion, consider the forces acting perpendicular to the ski track in the plane of the slope. The same conditions apply to the forces acting parallel to the ski track, but we will not consider those forces here. Assume that the ski moves parallel to the edge arc determined by the geometry we discussed in Technote 4, p. 205. There can be no skidding—no motion of any part of the ski edge perpendicular to the edge. The lateral position of the ski causes the snow reaction force to pass through the skier's center of mass. This configuration is illustrated in Chap. 4 in Figs. 4.11(a) and 4.11(b), p. 101, which shows a skier planted directly over his skis at the optimum tilt angle, ϕ, given by relation (4.9d), p. 102. The reaction force and the edging angle also depend on the turn radius through the centrifugal inertial force F_C.

The natural position of the plane of the ski is perpendicular to a line drawn from the skier's foot to the skier's center of mass, which is at about the hip. However, by using body angulation or by bending at the knee with lateral rotation, the skier can rotate his foot plane away from this natural position. Of course, no ski ever moves in the idealized manner we assume here; some skidding always occurs over some part, if not all, of the ski's edge. In general, skiers make turns with radii much smaller than those predicted by our model.

If no skidding occurs during a turn, the geometry of the ski when it is deflected to contact the slope determines the radius of that turn. A ski with a sidecut radius of R_{SC} set at an edge angle to the slope of Φ has a turn radius R_{con} equal to $R_{SC} \cos \Phi$, as we demonstrated in Technote 4, relation (T4.3), p. 206. In actual skiing, centrifugal force, which is determined by the skier's velocity and the turn radius, combines with the force component of the skier's weight W, which is parallel to the slope but perpendicular to the ski track. These two forces taken together set the tilt angle ϕ, so let us set the expressions for the equilibrium tilt angle ϕ_r and the edge angle Φ equal in order to determine the turn radius that will be consistent with both expressions. The tilt angle is given in Chap. 4 by the mathematical expression (4.9d), p. 102, which we repeat here as

$$\tan \phi_r = \frac{v^2}{g R_{con} \cos \alpha} - \tan \alpha \cos \beta. \qquad (T5.1)$$

Now, we replace the value of R_{con} in this expression with the geometric value for the turn radius that we worked out in Technote 4:

$$R_{con} = R_{SC} \cos \phi_r \qquad (T5.2)$$

The equilibrium tilt angle ϕ_r must satisfy relation (T5.3) given below, in which α is the slope angle and β is the track angle on the slope measured from the horizontal in the uphill quadrant of a turn [see Chap. 4, Figs. 4.5(a) and 4.5(b), p. 90]. Because the track angle β becomes greater than 90° as the skier passes the fall line, the second term in relation (T5.3) becomes positive in the downhill quadrant of a turn:

$$\tan \phi_r = \frac{v^2}{g R_{SC} \cos \phi_r \cos \alpha} - \tan \alpha \cos \beta. \qquad (T5.3)$$

We cannot solve this relation directly for ϕ_r, but the method of successive approximations yields a solution easily from which, by relation (T5.2), we may calculate the turn radius.

Howe works out an alternative method of calculation that generates a relation for the carved turn radius R_{con} that eliminates ϕ_r [1]. Howe starts by assuming that the skier is in dynamic equilibrium, such as the skier illustrated in Chap. 4, Fig. 4.11, p. 101, so that $F_N/F_{load} = \cos \phi_r$. Recall that F_N is the force normal, or perpendicular, to the slope and that F_{load} is equal and opposite to the snow-reaction force on the bottom of the skis. In the geometric relation given in Eq. (T5.2), the edge angle of the ski, Φ, is set to the dynamic equilibrium value, so $R_{con}/R_{SC} = \cos \phi_r$. From this, the relation $R_{con}F_{reac} = R_{SC}F_N$ follows. From the expressions worked out in Chap. 4 to describe the skier making the turn illustrated in Fig. 4.11 [Eqs. (4.9a)–(4.9d), pp. 101–102], we can write

$$F_{tl} = F_C - F_{lat}, \quad F_C = Wv^2/gR_T, \quad F_{lat} = W \sin \alpha \cos \beta,$$

$$F_N = W \cos \alpha, \quad F_{load}^2 = (F_C - F_{lat})^2 + F_N^2.$$

From these relations, the expression given below in Eq. (T5.4) follows:

$$F_N^2 R_{SC}^2 = F_{load}^2 R_T^2 = F_{load}^2 R_{con}^2,$$

since $R_T = R_{con}$,

$$[(F_C - F_{lat})^2 + F_N^2]R_{con}^2 - F_N^2 R_{SC}^2 = 0,$$

$$[F_C^2 - 2F_C F_{lat} + F_{lat}^2 + F_N^2]R_{con}^2 - F_N^2 R_{SC}^2 = 0. \tag{T5.4}$$

Now, insert the expression for the centrifugal force F_C to obtain the following quadratic equation for R_{con}:

$$(F_{lat}^2 + F_N^2)R_T^2 - 2\frac{v^2 W}{g}F_{lat}R_T + \left(\frac{v^2 W}{g}\right)^2 - R_{SC}^2 F_N^2 = 0. \tag{T5.5}$$

Now, when we insert the expressions for F_{lat} and F_N, which were given above in our discussion of the skier making the turn illustrated in Fig. 4.11 [Eqs. (4.9a)–(4.9d), pp. 101–102], we get

$$aR_T^2 + bR_T + c = 0,$$

where the coefficients become

$$a = [(\sin \alpha \cos \beta)^2 + \cos^2 \alpha],$$

$$b = -2 \frac{v^2}{g} \sin \alpha \cos \beta \qquad \text{(T5.6)}$$

$$c = \left[\left(\frac{v^2}{g} \right)^2 - R_{SC}^2 \cos^2 \alpha \right].$$

The solution for the quadratic expression above is

$$R_T = \frac{-b \pm \sqrt{(b^2 - 4ac)}}{2a}, \qquad \text{(T5.7)}$$

where in the uphill quadrants of the turn when $\beta < 90°$, the b term is negative. In the downhill quadrants, $\beta > 90°$, so $\cos \beta$ changes sign and the b term becomes positive. Only the positive square-root term gives physical solutions, and the term under the square root will be positive for physical solutions. Thus R_{con} for the turn before crossing the fall line is always greater than the radii generated for completion of a turn or after passing the fall line.

It takes some careful analysis to sort out the solutions given by Eq. (T5.7). R_{con} must always be less than R_{SC}, and if the sign of R_{con} changes, that implies that the ski is rolled from the inside edge to the opposite edge. For the case that $R_{con} = R_{SC}$, or when the ski rides flat on the snow bed, the total transverse force vector, F_{tl}, is zero. From the relation (4.9b) given in Chap. 4, p. 101, the relation for the critical traverse angle β_c follows:

$$\cos \beta_c = \frac{v^2}{g R_{SC} \sin \alpha}. \qquad \text{(T5.8)}$$

The calculation of the turning radius R_{con} does not give us much insight regarding what actually happens during a carved turn. The approach that generates the equilibrium tilt angle probably illuminates a bit more clearly what actually happens during the carved turn, so we use that approach in a series of calculations below to consider some illustrative cases.

Before showing some results from the actual calculation, we should understand the role of the critical traverse angle, β_c, calculated in relation (T5.8). Consider the example shown in Fig. 4.10, p. 100, in which a skier at the start of a turn [position (1)] generates a centrifugal force F_C less than the lateral force F_{lat}, so the skier must lean outward from the center of the turn, or uphill, by 8° to achieve dynamic stability. The skier cannot carve downhill or with the center of the turn to the right without using body angulation. At point (2) in the turn, there is no lateral force acting, so the skier is precisely over the ski. At point (3) on the fall line the skier is subject to centrifugal force F_C only, so the skier leans inward toward the center of the turn by 23°. The inward lean increases to 45° at the completion of the turn. At every point in this turn the skier's skis are carving at an angle of 23° to ensure the constant-radius turn, but the skier rolls the skis relative to his body by 23°+8°, or 31° to his right at the start, to 45°−23°, or 22° to his left, at the completion of the turn to maintain a constant turn radius.

In setting up the general problem, we assume that the skier turns to the right or downhill from his traverse. Relation (T5.8) given above defines a ski inclination angle at which a skier must track in the uphill quadrant in order to ride on the inside ski edge as the velocity demands. The examples we will use are graphed in Fig. T5.1. The critical traverse angle is shown as

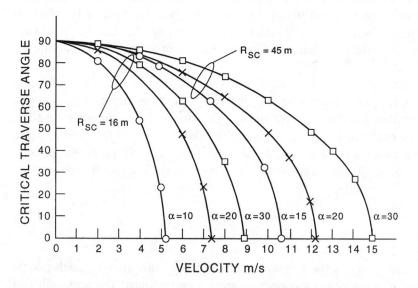

FIGURE T5.1. Critical traverse angle vs velocity for carving toward the center of the turn. The edge angle Φ becomes 0° at the critical transverse angle; thus $R = R_{SC}$. Carving toward the center of the turn is possible for velocities greater than those shown by the given curve.

a function of velocity for several slope angles α, given in degrees. Two models of skis are used: a conventional ski with a sidecut radius R_{SC} equal to 45 m, and a high-performance slalom or sport ski with a sidecut radius of 16 m. These curves assume that the skier provides no body angulation to change the edging angle of the skis. The velocities to the right of each curve provide for carving on the inside edge of a right turn. The velocities to the left of each curve represent cases for which the outside edge of the ski will carve and the turn will change to a left turn. Remember that a traverse angle of 90° represents a track down the fall line. These curves illustrate the difficulty of initiating a carved turn at small traverse angles without using body angulation or deliberate ski rotation and placement.

Let us now consider some calculations for purely carved turns in which the inside edge of the ski can carve the full 180° turn from a horizontal traverse in one direction at the start to a horizontal traverse in the opposite direction at the finish. From Eq. (T5.3), calculations for the equilibrium tilt angle ϕ_r were solved by numerical approximation. Once values for ϕ_r as a function of β are available, we may convert them to values of R_{con} by using Eq. (T5.2). The results of this conversion are graphed in Fig. T5.2. For

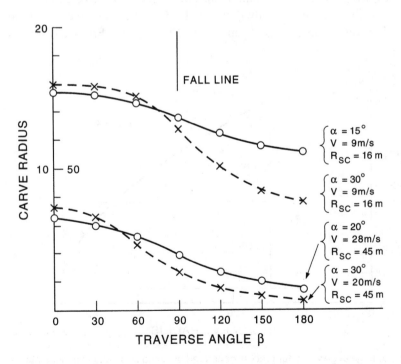

FIGURE T5.2. Carved turn radii versus the slope angle α and the traverse angle β.

simplicity, we assume constant velocities, which, it turns out, is not far from experience in field measurements, because if velocity were to increase throughout a turn, the turn radii would be smaller than those indicated by the curves with an increase in traverse angle.

For the ski with a 16-m R_{SC}, we assume a velocity of 9 m/s, or about 30 ft/s (20 mph), which might represent the speed of a slalom racer. For the ski with a 45-m R_{SC}, we assume a velocity of 20 m/s or 66 ft/s (45 mph), which might represent the speed of a giant slalom racer. The slope angles range from gentle to modest. The left-hand scale for the "carve radius" applies to the ski with R_{SC}=16 m; the right-hand scale applies to the ski with R_{SC}=45 m. These same results could also be obtained by using the quadratic relation for R given above in Eq. (T5.7). The equilibrium tilt angle of the R_{SC}=45 m ski, ϕ_r, grows to about 80° at the end of the turn. That value is probably excessive; the snow surface could not continue to hold the ski's edge at that angle to the slope. For the R_{SC}=16 m ski, the maximum inclination for ϕ_r is about 54°, which is a reasonable value for actual skiing.

Finally, consider how skiers use body angulation to change the effective edging angle of their skis. From the relation given in Eq. (T5.2), we calculated the carve radius, R_{con}, for a giant slalom ski with a 45-m R_{SC} with its edges set initially at 35° and then, using body angulation, changed by 30° up or down. The results are shown in Fig. T5.3. Substantial changes in the carve radius result in a range from about 20 to 45 m. From the results

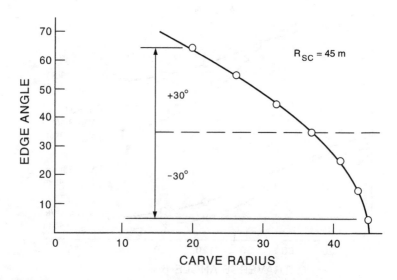

FIGURE T5.3. Change in the turn radius generated by using body angulation to change the edge angle.

presented in Fig. T5.3, it appears to be inevitable that carved turn radii become shorter throughout a turn and that the equilibrium body tilt must increase, whether the ski edge angle increases or not. Thus we see that during the carving of a turn the pressure the ski edge exerts on the snow increases greatly, to the point that the snow may not be able to hold the edge. If the edge breaks free, the ski skids out, and usually the skier falls.

REFERENCE

1. Much of this discussion is indebted to J. Howe's work presented in *Skiing Mechanics* (Poudre, LaPorte, CO, 1983).

UP-AND-DOWN UNWEIGHTING

An analytical understanding of the forces involved in the process of unweighting skis through up-and-down body motion should help skiers appreciate what professional ski instructors mean when they advise skiers to make an "up–turn–down" or, as is more likely with modern ski instruction, a "down–turn–up" sequence through a carved turn. For the purposes of our discussion, we will assume that the vertical displacement of a skier's center of mass caused by the up-and-down body motion is 14 in. The actual range of possible vertical displacements is probably from about 1–2 ft.

Refer to Fig. 5.1, p. 110, and assume that the skier illustrated descending the slope is skiing on a plane surface. The equation of motion of the skier's center of mass normal to the plane is

$$\frac{W\mathbf{a}}{g} = \mathbf{F}_{reac} - \mathbf{F}_N. \qquad (T6.1)$$

This is the same expression given in Chap. 5 as Eq. (5.1). Set $\mathbf{F}_N = W \cos \alpha$, where α is the slope angle. If the snow-reaction force, $\mathbf{F}_{reac} > \mathbf{F}_N$, the acceleration a is positive, or upward, and depends on the magnitude of $\mathbf{F}_{reac}$. If $\mathbf{F}_{reac} < \mathbf{F}_N$, the acceleration a is negative, or downward, and is limited by $g \cos \alpha$ because the skier is in free fall.

From the equations of motion, we can calculate the skier's vertical velocity v and vertical displacement s. Using a simple case for our example, assume that the skier's velocity down the slope is 20 mph, or about 30 ft/s, and that the skier's motion is on a horizontal plane. The downward acceleration g will be generated by the force of gravity W when the skier's legs are retracted. Assume that the thrust for upward motion, F_{reac} is $2W$, so the upward acceleration is also g. For the velocity and

216

displacement of the initial down motion, the equation of motion given above in Eq. (T6.1) gives us,

$$v = -gt, \quad \rightarrow h = -\tfrac{1}{2}gt^2 \tag{T6.2}$$

For the up motion that follows in this sequence, the equations of motion for velocity and displacement become

$$v = -v_d + gt \rightarrow h = -h_d - v_d t + \tfrac{1}{2}gt^2, \tag{T6.3}$$

where $-v_d$ is the skier's velocity at the end of the down acceleration phase and $-h_d$ is the skier's position. This assumes that the skier's leg retraction lasts for 0.2 s until the skis touch down. At that point to stop the skier's fall, the skier's legs are further compressed with an upward acceleration off the surface of the snow, which lasts 0.2 s, which is again equal to g.

Now refer to Fig. 5.4(a), p. 117, where the velocity v and the displacement h are plotted for a skier executing a down–up sequence, unweighting maneuver. Note that 0.2 s is about the shortest possible time for the skier's body to respond, and the distance the skier's body falls during the unweighting maneuver is $h=0.64$ ft. The skier must also perform the desired up maneuver because a weighting occurs immediately. For every thrust there must be an equal and opposite, time-integrated, complementary thrust because the time integral of the thrust represents the change in momentum and hence in velocity. Every down–up motion results in a final velocity of 0: the downward velocity grows to -6.4 ft/s, during the up thrust, it returns to 0, but the body moves downwards another 0.64 ft. The total displacement is 1.28 ft downward. To recover his original, vertical position, the skier would have to perform an up–down sequence at some later time. For the sake of simplicity, we assume that the forces given in this example are constant during the two intervals. In this down–up maneuver, the unweighting occurred first, so the skier could either set the edges of his skis in a different direction or rotate them.

The up–down sequence, unweighting maneuver is shown in Fig. 5.4(b), p. 117. The initiating up thrust sets the skier's center of mass in vertical motion, giving us the following expressions for velocity and displacement, respectively:

$$v = gt \rightarrow h = \tfrac{1}{2}gt^2. \tag{T6.4}$$

The velocity and displacement after the up phase are 6.4 ft/s and 0.64 ft, respectively. The down phase gives the following expression:

$$v = v_u - gt \rightarrow h = h_u + v_u t - \tfrac{1}{2}gt^2. \tag{T6.5}$$

When the velocity becomes 0, the skier's upward displacement is 1.28 ft at the time of 0.4 s. Subsequently, the skier must perform a down cycle to return to his original, crouched position. If this subsequent down cycle occurs immediately after the initial up cycle, the skier can achieve an unweighted time out to the point indicated by 0.4 s. For the up–down sequence, unweighting is delayed to 0.2 s. Eventually an up thrust must occur to bring the body to rest in a crouched position, ready for the next leg thrust cycle.

Time is shown in Figs. 5.4(a) and 5.4(b), p. 117, in intervals of 0.2 s. The skier travels a horizontal distance d at a velocity v of 30 ft/s. During the unweighting action illustrated in Fig. 5.4(a) the skier travels about a ski length; for the unweighting action illustrated in Fig. 5.4(b), the skier travels about two ski lengths. These distances are borne out by field observations.

ANALYSIS OF PREJUMPING

When confronted by a break in the slope on the track they wish to ski, skiers may reduce the time they spend airborne by prejumping the break. The following calculations quantitatively demonstrate the effect of such prejumping. Refer to Fig. 5.5, p. 119, which illustrates the slope break we will take for our example.

A skier who does not prejump the break lands at a distance d_l from the break, C. A skier who prejumps the break to land at C travels the distance 2 dm, which we will see is equal to d_1. The horizontal track is followed by the sloping track at the angle δ. On a slope of angle α, replace the acceleration of gravity with $g \cos \alpha$, and neglect the acceleration perpendicular to the slope. The initial horizontal velocity is v. The variable h measures the vertical distance from the initial plane. A skier running off the track will travel in the horizontal and vertical directions for the same time t. Hence

$$h = \tfrac{1}{2}gt^2 = \tfrac{1}{2}g\left(\frac{d_l}{v}\right)^2 = d_l \tan \delta, \quad \text{since } d_l = vt, \tag{T7.1}$$

$$d_1 = \frac{2v^2}{g} \tan \delta = 2d_m. \tag{T7.2}$$

One may prejump to the left of the break point in the slope C to land tangent to the slope or with a vertical velocity component of $v_{h0} = v \tan \delta$. Here v_{h0} also equals the vertical take off velocity at the distance d to the left of C. The flight time t and the distance d are given by

$$t = \frac{2v_{h0}}{g} \rightarrow d = vt = \frac{2vv_{h0}}{g} = \frac{2v^2}{g} \tan \delta = 2d_m \tag{T7.3}$$

Prejumping to land at the vertex C with a velocity parallel to the slope gives the same flight distance as that yielded by the horizontal takeoff, provided the skier tucks again after the takeoff to assume the initial body configuration.

Can the flight distance be shortened? Assume the takeoff point T is at a distance d_T to the left of C and that the skier has the same horizontal velocity as that given above and that the skier's legs and skis are tucked an amount h. The skier falls the amount h as the skier grazes the break point on the slope and continues to fall until landing. At the break point in the slope, C, the skier's vertical velocity becomes $v_{h0}=gt=gd_T/v$. The height h for the trajectory to pass through C is the fall distance for the same time:

$$h=\tfrac{1}{2}gt^2=\tfrac{1}{2}g\left(\frac{d_T}{v}\right)^2 \rightarrow d_T=\left(\frac{2h}{g}\right)^{1/2}v. \tag{T7.4}$$

The skier lands at a distance d_L to the right of C, where $t=d_L/v$, and $y=d_L \tan \delta$ is given by

$$y=\frac{v_{h0}d_L}{v}+\tfrac{1}{2}g\left(\frac{d_L}{v}\right)^2=d_L \tan \delta,$$

$$d_L=\left(\tan \delta-\frac{v_{h0}}{v}\right)\frac{2v^2}{g}. \tag{T7.5}$$

The total horizontal flight distance is $d_h=d_T+d_L$, which gives us

$$d_h=\frac{2v^2}{g}\tan \delta-\frac{2vv_{h0}}{g}+\left(\frac{2h}{g}\right)^{1/2}v. \tag{T7.6}$$

Now $v_{h0}=(2gh)^{1/2}$, so

$$d_h=\frac{2v^2}{g}\tan \delta-\left(\frac{2h}{g}\right)^{1/2}2v+\left(\frac{2h}{g}\right)^{1/2}v \rightarrow d_h=\frac{2v^2}{g}\tan \delta-\left(\frac{2h}{g}\right)^{1/2}v. \tag{T7.7}$$

This flight trajectory lands the skier parallel to the slope when $v_{h0}/v=\tan \delta$, or $d_L=0$. If h_m is the height for horizontal release to make contact at C, $\tan \delta=(2gh_m)^{1/2}/v$. Then d_h becomes

$$d_h=2\left(\frac{2h_m}{g}\right)^{1/2}v-\left(\frac{2h}{g}\right)^{1/2}v. \tag{T7.8}$$

Set $d_m=(2h_m/g)^{1/2}v$, and because $d_T=(2h/g)^{1/2}v$, write d_h as

$$d_h=2d_m-d_T. \tag{T7.9}$$

The optimal distance for the skier to prejump the slope break in a tucked position would yield a flight trajectory that just touches at the break. As h increases, the distance d_T must increase. Note that $2d_m$ is exactly the distance at which to prejump and land at C parallel to the slope. If one could prejump by tucking at the height h_m, then $d_h = d_m$ is the minimum flight distance.

Note that launching with v_y upward always extends the trajectory; it is impossible to launch with v_y downward, which would shorten the trajectory. Of course, after passing the break in the slope at C, the skier could extend his skis downward to shorten his flight distance and be in a body configuration that would allow the skier to absorb the landing impulse.

AERODYNAMIC DRAG

The aerodynamic drag force on a skier resulting from the viscosity of the air consists of two different types. If air flows in a perfect, streamlined, laminar flow around the skier's body, the aerodynamic drag on the skier results from the shear velocity gradient at the surface and increases or decreases proportionally with the velocity of the air flow. Drag from laminar air flow is negligible in the context of this discussion. As the velocity of the air flow around the skier increases, however, the shear also increases until the laminar flow separates at some angle ϕ on the downstream side of the skier's body and turbulence develops in the wake that trails behind the skier. The momentum that transfers from the skier to the air in that turbulent wake represents a drag force on the skier.

The relation that describes this aerodynamic drag phenomenon is similar to the relation that describes plowing through any substance; the drag that results on an object when it plows through the air is given by the difference in pressure between the front and trailing sides of the object. Plowing, or pressure drag, is the significant drag factor that affects a skier's performance, and it has been studied in connection with a variety sports [1].

A simple, dimensional analysis of the general form of the drag force on an object follows [2]. The variables upon which the drag force depends are density ρ, viscosity μ, the dimension of the body l, and velocity v. By comparing the dimension of force with any function of all of these variables, the relation must take the form shown below:

$$F_D = \tfrac{1}{2}\rho A C_D v^2. \qquad (T8.1)$$

In this relation, A is the frontal area of the body in the air flow, having a dimension of l^2, and C_D is the drag coefficient, which is determined by the Reynolds number. The dimensionless Reynolds number determines the conditions for transition from laminar airflow to turbulent airflow around the body under study, and thus it correlates with the magnitude of the drag coefficient C_D. The Reynolds number is given by $Re = lv/\nu$, where l is a dimension of the body in the air flow, v is the velocity, and $\nu = \mu/\rho$, the kinematic viscosity of air, equal to 1.3×10^{-5} m^2 s^{-1} at 1 atm of pressure and 0 °C.

Let us look more closely at the drag effects created by the transition from laminar to turbulent airflow as it affects the spherical body illustrated in Fig. T8.1. Figure T8.1(a) shows a perfectly streamlined, laminar airflow that results in no separation of the flow lines once they pass around the sphere. Note that the flow line that encounters the sphere at point F resumes immediately at point R after the air flows over the surface of the sphere. The second and third cases illustrate circumstances in which the airflow velocity determines the Reynolds number at which the laminar airflow lines separate, creating two differently sized areas of turbulence. The greater pressure drag occurs in Fig. T8.1(b).

The airflow in these examples acts like a low-viscosity fluid, so Bernoulli's theorem for nonviscous fluid flow describes the phenomenon. Bernoulli's theorem states that $P + \frac{1}{2}\rho v^2 = B$, where B is a constant. The term $\frac{1}{2}\rho v^2$ is the kinetic energy per unit volume of the fluid; P is the local

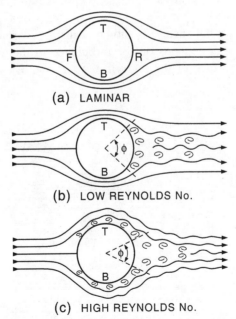

(a) LAMINAR

(b) LOW REYNOLDS No.

(c) HIGH REYNOLDS No.

FIGURE T8.1. Three different airflows around a sphere: (a) laminar flow; (b) low Reynolds number with early separation of the airflow; (c) high Reynolds number showing delayed separation of the airflow.

pressure in the fluid. Because of its viscosity, the energy per unit volume, or B, decreases for the fluid near the surface of the sphere as the fluid moves downstream. Thus, for the sphere illustrated in Fig. T8.1(a), the kinetic energy at points T or B is sufficient to allow the airflow at the surface of the sphere to move freely against the higher pressure downstream to the point R. Because the pressure is the same at points F and R, there is no pressure drag force on the sphere. For the case illustrated in Fig. T8.1(b), however, the kinetic energy at T or B is reduced. It is only sufficient to drive the fluid mass to the points indicated by the separation angle ϕ because the constant B in Bernoulli's equation has been reduced. This implies that the stagnation pressure in the turbulent wake behind the sphere is lower than the pressure over the forward part of the sphere, which results in a pressure drag force that pushes the sphere in the direction of the fluid motion.

Now look at what happens when a turbulent surface flow sets in upstream of the point at which the laminar airflows separate, as it does in Fig. T8.1(c). When the airflow is turbulent ahead of the separation point, the flow velocity near the surface of the sphere is larger than it is for a laminar airflow, so the kinetic energy term $\frac{1}{2}\rho v^2$ is larger, and the stagnation point moves downstream to a smaller angle ϕ and the pressure drag is reduced correspondingly. The pressure drag is given by Eq. (T8.1), and the value of the drag coefficient C_D, varies with the Reynolds number.

Let us now consider Fig. T8.2 where the transition from laminar to turbulent flow over a flat surface is shown occurring over a certain distance l_{crit}. The value of l_{crit} depends on the turbulence in the airstream and the roughness of the surface of the body in the airflow. The velocity profiles

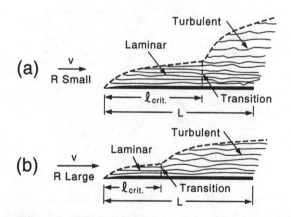

FIGURE T8.2. Effect of the Reynolds number on the transition from laminar to turbulent flow (Millikan, 1941).

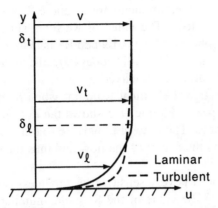

FIGURE T8.3. Velocity profiles for laminar and turbulent flow near a surface (Millikan, 1941).

adjacent to the surface of the body in the airflow are shown in Fig. T8.3. The drag force due to the velocity gradient is lower for the laminar flow than it is for the turbulent flow, but the contribution to the total drag force made by either surface laminar or turbulent flow is usually much smaller than the force that results from plowing or pressure drag.

At a critical value of the Reynolds number, Re, the laminar flow regime changes into a surface boundary layer of turbulent flow. The relevant questions are, at what values of Re does the transition occur, and how much is the subsequent change in C_D, the drag coefficient? Figure T8.4 shows the

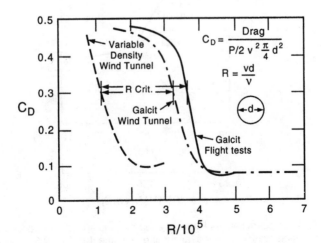

FIGURE T8.4. Wind-tunnel measurements of the drag coefficient C_D for a sphere (Millikan, 1941).

results of wind tunnel measurements for a sphere. Notice that the values of $R_{crit}/10^5$ vary from 1 to 4. The value of Re for a skier at $v = 10$ m s^{-1} is about 2×10^5. For spheres, $C_D = 0.45$ when Re is below 1.4×10^5; for larger values of Re, the drag coefficient is about 0.07. Thus the drag drops markedly at the critical value of velocity.

A skier goes through the critical Re point when he moves from low to modest downhill speeds. Figure T8.5 shows the dependence of C_D on Re for a variety of shapes. The ratio of transverse to longitudinal dimension of the body determines the nature of the flow and thus the drag coefficient for that body.

The critical airspeed is very sensitive to roughness on the surface of the body in the airflow. The value of the Reynolds number at which the drag

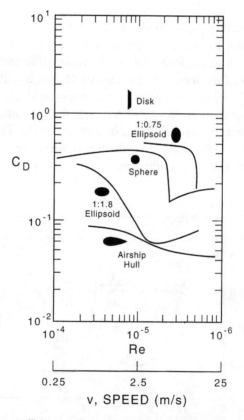

FIGURE T8.5. Drag coefficient vs Reynolds number and speed for bodies with different shapes. The speed was calculated assuming a value for l of 0.5 m and for v of 1.25×10^{-5} m^2 s^{-1}. Note the change in behavior with the length of the object (Rouse and Howe, 1953).

TABLE T8.1. *Skier drag forces (newtons).* $m=90$ kg, $A=0.3$ m^2; and $\rho=1.0$ kg m^{-3}.

Slope angle (deg)	Friction ($\mu Mg \cos \phi$) Coefficient μ			Air drag ($\frac{1}{2}C_D\rho A v^2$) Velocity (m s^{-1})					
	0.02	0.05	0.08	10	20	30	10	20	30
				$C_D=0.07$			$C_D=0.45$		
15	17.0	42.6	68.0	1	4.2	9.5	6.8	27	61
30	15.3	38.2	61.2	1	4.2	9.5	6.8	27	61
45	12.5	31.2	50.0	1	4.2	9.5	6.8	27	61

factor drops may be lowered by a factor of 3 for a body with a rough surface. For example, golf balls and footballs have puckered surfaces that affect not only the distances these spinning balls fly when they are hit or thrown but the curves of their trajectories as well.

There is not much in the literature on the aerodynamics of skiing regarding the use of ski suits or helmets with roughened surfaces [3]. This is odd, considering that if puckering or dimpling a skier's suit or helmet might reduce the air-pressure drag on the skier, the total drag force on the skier would be reduced considerably. To put in context the relative amounts of drag exerted by snow friction and by aerodynamic forces, look at Table T8.1, which gives values for each of the drag forces under different parameters. For the likely friction coefficients, $\mu<0.05$, and at the greater velocities, the air drag force is equal to or larger than the snow friction drag.

Roberts has analyzed aerodynamic drag on skiers using cylinders to model the skier's torso, head, arms, and legs [4]. In this study, observations for the drag coefficient C_D drop from 1.2 to 0.3 at a Reynolds number between 200 000 and 400 000. The actual transition Reynolds number probably depends on the configuration of the object under study, and thus it would change as a skier changes his stance.

Given the importance of aerodynamic drag as a factor in slowing a skier's descent of the hill, surely some research directed specifically at the effect on drag resulting from having rough or dimpled surfaces on the skier's helmet and clothing is warranted.

REFERENCES

1. See, for example, N. de Mestre, *The Mathematics of Projectiles in Sports* (Cambridge University Press, Cambridge, U.K., 1990).
2. The discussion of aerodynamic drag that follows is indebted to C. B. Millikan's analysis of aerodynamics, *Aerodynamics of the Airplane* (Wiley, New York, 1941); for the parts of the discussion more specifically centered on skiing, we are indebted to S. C.

Colbeck's "An Error Analysis Of The Techniques Used In The Measurement Of High-Speed Friction On Snow," Ann. Glaciol. **19**, 19 (1994), and M. S. Holden's "The Aerodynamics of Skiing," Sci. Am. **258** (2), T4 (1988).

3. Both Holden (1988), cited in Ref. 2, and A. E. Raine, "Aerodynamics of Skiing," Sci. J. **6** (3), 26 (1970), present the results of wind-tunnel tests of the air drag on speed skier suits, boots, and fairings, as well as the body positions a skier might assume. Neither author mentions tests of rough-surfaced or dimpled equipment or clothing.

4. C. C. Roberts Jr., "Numerical Modeling of the Transient Dynamics of a Skier While Gliding," in *Biomechanics of Sport: A 1987 Update*, edited by E. D. Rekow, J. G. Thacker, and A. G. Erdman (American Society of Mechanical Engineers, 1987).

THE BRACHISTOCRONE PROBLEM
The Path of Quickest Descent

In ski racing, the objective is to minimize the time of descent through a prescribed course delineated by poles, or gates, around which the skier must turn. Segments of the course consist of direct paths from one gate to the next, which are usually placed in a regular array at some distance down and across the slope from each other. Finding the path of quickest descent between such pairs of gates is not done simply by following a straight line between them. This problem, known as the brachistochrone problem, requires finding the stationary value of the integral of a function for different paths between two fixed points. It is treated by the calculus of variations and is discussed in any of a number of texts on mathematical physics or mechanics [1].

Refer to Figure T9.1, which illustrates the case for a skier who starts at rest from the origin and wishes to minimize his transit time to some point B, which lies down and across the slope [2]. The figure shows the skier's trajectory in the plane of the slope. Given that the skier starts at rest, the length of an element of the skier's path along the curve (ds) and the skier's velocity (v) at any point downslope are as follows:

$$ds = \sqrt{1 + y'^2}\,dx, \quad y' = \frac{dy}{dx}, \quad v = \sqrt{2g'y}, \tag{T9.1}$$

where $g' = g \sin \alpha$ is the effective gravitational acceleration for the slope. The effects of friction and drag are neglected in this example. Therefore the time t required to move from the origin to point B is

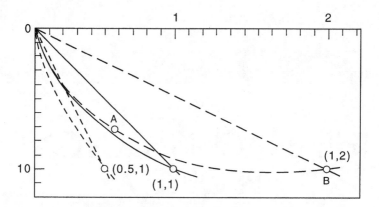

FIGURE T9.1. Brachistochrone trajectories on an inclined plane. Three endpoints (x,y) of $(0.5,1)$, $(1,1)$, and $(1,2)$ are shown. The curves are independent of the acceleration of gravity on the plane and hence the slope. As x of the end point decreases, the trajectory approaches that of a straight line. In each case the mass starts at rest from the origin.

$$t=\int_0^{y_0}\sqrt{\frac{1+y'^2}{2g'y}}\,dx. \tag{T9.2}$$

The integrand does not depend on x, so the solution of the variational problem leads to the relations given below.

$$F-y'\frac{\partial F}{\partial y'}=\text{const}, \quad F=\sqrt{\frac{1+y'^2}{2g'y}} \tag{T9.3}$$

The relations in (T9.3) give, in turn, the following relations:

$$y(1+y'^2)=C. \tag{T9.4}$$

If $y'=\cos\theta$, then

$$y=\frac{C}{1+y'^2}=C\sin^2\theta=\frac{C}{2}(1-\cos 2\theta), \tag{T9.5}$$

$$\frac{dx}{d\theta} = \frac{1}{y'}\frac{dy}{d\theta} = C \tan \theta \sin 2\theta = C(1 - \cos 2\theta). \qquad \text{(T9.6)}$$

Integrating Eq. (T9.6) and substituting $C = 2A$ and $2\theta = \phi$, we get

$$x = \frac{A}{2g'} (\phi - \sin \phi),$$

$$\text{(T9.7)}$$

$$y = \frac{A}{2g'} (1 - \cos \phi).$$

This is the parametric equation of a cycloid, or the curve generated by a point on a wheel of a radius $R = A/2g'$ as it rolls along a straight line. Since g' is outside the integral in the transit time expression, the shape of the trajectory is independent of the acceleration. The curve is shown in Fig. T9.1 for the end points generated by the values (0.5,1), (1,1), and (2,1), respectively. Remember, in each of these cases, the skier starts from rest at the origin.

A skier who starts from some point downslope, such as point A in Fig. T9.1, with some initial velocity follows the same curve as the skier who starts from an origin at rest, but he starts some distance down the plane, $y_A = v_A^2/2g'$, at the appropriate coordinate, x_A. In this case, the value of y_A and the coordinate separations of x and y between the initial and final points determine ϕ_A, ϕ_B, and R, which give the appropriate coordinate x_A.

Using Eq. (T9.7), let us consider the case for the track shown in Fig. T9.1 that illustrates the (1,1), that is, $x_B = y_B$, example. The time for the straight track shown in the figure is the distance traveled divided by one-half of the final velocity because the acceleration is constant. The times for the straight line and the cycloid trajectories become

$$t = \begin{cases} 2\sqrt{\dfrac{y_B}{g'}}, & \text{straight} \\[3ex] 1.82\sqrt{\dfrac{y_B}{g'}}, & \text{cycloid} \end{cases} \qquad \text{(T9.8)}$$

The time difference increases as the separation between the straight line and the cycloid curve increases. Conversely, as the skier starts with a larger initial velocity, the separation between the two lines decreases and the subsequent difference in time between the straight track and the minimum time track also decreases.

REFERENCES

1. For example, see J. Mathews and R. L. Walker, *Mathematical Methods of Physics* (W. A. Benjamin, New York, 1964), p. 307. The discussion here of the brachistochrone problem in the context of skiing is generally indebted to G. Twardokens, *Universal Ski Techniques* (Surprisingly Well, Reno, NV, 1992), pp. 256–257.
2. The details of the mathematical development sketched here may be found in G. Reinisch, "A Physical Theory of Alpine Ski Racing," Spektrum Sportwissenschaft **I**, 27 (1991).

PUMPING TO INCREASE VELOCITY

Skiers can increase their velocity during a downhill run on an undulating slope by using body motions to increase their kinetic energy [1]. In a different context, an example of this phenomenon is the figure skater who, in a leaping pirouette, brings her outstretched arms in closer to her body, which increases her angular velocity and, hence, the kinetic energy of her system, so she accelerates through the pirouette in midair without ever touching the ice. So too a skier may increase his kinetic energy and acceleration during a turn or when he runs the troughs in a mogul field by pumping up and down with his legs to shift his center of mass. In effect, there is a reservoir of potential energy in the human body that the skier may convert to kinetic energy. We can analyze this energy-conversion process by using the physical principles associated with work and energy.

Consider the cart with a human rider shown below in Figure T10.1. For the purposes of our example, assume that the cart is given an initial horizontal velocity and that it moves without friction along the curved track in the illustration. The rider can pump himself and the cart to a higher velocity if anywhere in his transit of a curve he pushes his center of mass toward the center of that curve. When the rider moves his center of mass in this fashion, his velocity must increase because the kinetic energy for rotation is given by the relation

$$T = \frac{L^2}{2I} \tag{T10.1}$$

where T is the kinetic energy, L ($=I\omega$) is the angular momentum, and I is the moment of inertia about the axis of rotation. If we designate ω as angular velocity, the angular momentum equals the moment of inertia about the axis of rotation multiplied by the angular velocity: $L = I\omega$.

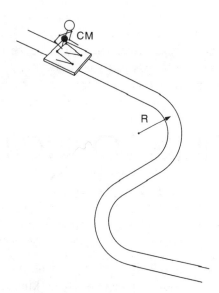

FIGURE T10.1. A person kneels on a cart as moves along a track on a horizontal floor through a curve with the radius R.

   Now, consider that the angular momentum of the cart and its rider cannot change if no external torques act on the system. Any reaction forces from the imaginary frictionless rails that hold the cart and its rider on its curved track are always at right angles to the motion, so no work can be done on the system from the outside. Note that $I \approx MR^2$, where M is the mass and R is the radius of the center of mass from the axis of rotation. Now, the rider can decrease the moment of inertia I by moving his body toward the center of rotation during the first arc. When he does this, the angular momentum L remains the same because no torque acts on the system, so the kinetic energy must increase. During the next arc, the rider can move his body again toward the center of rotation and in the same manner increase the kinetic energy and hence the velocity of the cart and rider system.

   When the rider moves his mass toward the center of rotation on each turn, work is done against the centrifugal force F_C in an amount exactly equal to the increase in the kinetic energy T, as it was given in relation (T10.1). The centrifugal force F_c increases with velocity, so the work done against F_c, and thus the increase in kinetic energy T, increases with velocity. The relation for centrifugal force is $F_C = Mv^2/R$, so the work done in moving a distance h is

$$dT = -h\,\frac{Mv^2}{R} \rightarrow \frac{\Delta T}{T} = -\frac{2h}{R}, \tag{T10.2}$$

where dT is the change in kinetic energy. Since the kinetic energy T equals $Mv^2/2$, the change in the kinetic energy is $dT = Mv\,dv$. The angular momentum of the cart as it goes through a curve in the track is MVR, and it is constant; so $hv + R\,dv = 0$. Using this relation to eliminate dv from the expression $dT = Mv\,dv$, we get $dT = -hMv^2/R$. Note that this is the same expression as the one given in relation (T10.2); the sign of h is negative here because the center of mass of the rider and the cart moves toward the center of rotation.

We can apply these same calculations to describe what happens when a skier goes through a trough at a velocity v. At the point of highest curvature, he rises up as high as possible from his initial crouched position. When he performs this body motion—pumping himself up with his legs—his kinetic energy changes as a result of the displacement of his center of mass effected by the pumping motion. From the relation given Eq. (T10.2), observe that $dT/T = -2h/R$. If $R = 4m$, and $h = -0.5m$, then $dT/T = 0.25$, which is a substantial increase in kinetic energy. The centripetal acceleration, $a = v^2/R$, for a velocity $v = 10$ m/s is 2.5 times that of gravity, so the reaction force on a skier's legs when he performs this maneuver at this speed is 3.5 times his weight.

This example illustrates what occurs when a skier anticipates a run through a trough. The skier comes into the trough in a crouch with his center of mass low to avoid being pitched forward, and then he tends to rise up going through the low point, which increases his velocity coming out of the trough. A less experienced skier might be caught off guard by the unexpected acceleration, lose his balance, and fall. An expert skier will use this same effect to his advantage and pump his body up and down at appropriate places on a race course. This body motion increases the skier's speed going through rolls on the race course or during the carving of a turn, particularly at the end of a turn where the radius of curvature is shortest and the potential for increasing kinetic energy, and thus velocity, by doing work against the centrifugal force is greatest.

REFERENCE

1. This discussion is indebted to the work of C. D. Mote, Jr. and J. K. Louie, "Accelerations Induced by Body Motion in Snow Skiing," J. Sound Vibr. **88**, 107 (1982).

THE SKIER AS AN INVERTED PENDULUM

Let us model the skier on the slope as a system assumed to be a pendulum inverted on its pivot point as shown in Fig. T11.1 [1]. The ski acts as the pivot point, and it may move sideways or be fixed by edging; in either case the ski provides the necessary vertical and transverse forces to make it the pivot point in our model. The rigid connection between the leg, the boot, and the ski provides stability fore and aft. The reaction force R balances the weight Mg, mass times the acceleration of gravity; the lateral force at the pivot point F balances the inertial force, $M\ddot{y}$. The rotational motion of the skier is important, so we write the equation of motion as follows: the pivot point O is not fixed; the equation is therefore for rotation about the center of mass:

$$-I\ddot{\gamma}+Mgl\gamma-Fl=0. \qquad (T11.1)$$

I is the moment of inertia about the center of mass. The pivot point is driven by F to skid sideways; the force F also drives the center of mass sideways. The approximation is also made that the angle γ is small, so $\tan\gamma\approx\gamma$. If $F=0$, the solution is of the form:

$$\gamma(t)=\frac{\gamma_0}{2}\,(e^{pt}+e^{-pt})=\gamma_0\cosh pt, \quad \text{where } p=\sqrt{\frac{Mgl}{I}}$$

$$(T11.2)$$

This solution shows that the inverted pendulum will tilt over with a time governed by the value of p. If $F=k\gamma$, where k represents the strength of the correction force needed to restore the pendulum to its original position, the expression becomes

$$I\ddot{\gamma}+(k-Mg)l\gamma=0. \qquad (T11.3)$$

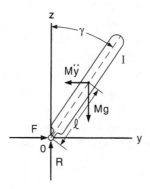

FIGURE T11.1. The geometry and forces on an inverted pendulum. [Reprinted with permission from J. M. Morawski, "Control Systems Approach to a Ski-Turn Analysis," J. Biomech. **6**, 267 (1973) (Elsevier Science Ltd., Oxford, England).]

With $k>Mg$, a solution starting with an initial tilt of γ_0 is oscillatory but stable. The expression for γ becomes

$$\gamma(t)=\gamma_0 \cos \omega t,$$

where (T11.4)

$$\omega=\sqrt{\frac{(k-Mg)l}{I}}.$$

The angle γ given in Eq. (T11.2) may increase in time without bound. Equation (T11.4) is an oscillatory solution. In practice, one wants a damped solution for which γ approaches the desired value, γ_0. So include a velocity dependent term in F; set $F=k\gamma+k_1\dot{\gamma}$, and the equation of motion is now

$$I\ddot{\gamma}+k_1l\dot{\gamma}+(k-Mg)\gamma=0. \qquad (T11.5)$$

Assume a solution of this equation to be $\gamma=e^{\kappa t}$. Then κ is given by the roots of Eq. (T11.5) as follows:

$$I\kappa^2+k_1l\kappa+(k-Mg)l=0,$$

where (T11.6)

$$\kappa=-\frac{k_1l}{2I}\pm\sqrt{\left(\frac{k_1l}{2I}\right)^2-\frac{(k-Mg)l}{I}}.$$

The solution is periodic when $k_1l/2I<[(k-Mg)L/I]^{1/2}$; then $\kappa=\kappa_0\pm i\omega$; otherwise the solution is unstable and grows exponentially.

Thus we see that the inverted pendulum by itself is obviously unstable, ready to fall to one side or the other at any instant. Consider, for example, how another example of an inverted pendulum, a pencil or a yardstick, will not stand on end at rest in the palm of one's hand; however, with the dynamical feedback introduced by visual and muscular response, most of us can balance a yardstick, even if only for a short time, in the palm of our hand. Doing the same for a pencil, with its much smaller moment, requires far greater skill. To see what this has to do with skiing, we need to look more closely at the mechanics of the inverted pendulum.

The period t of a stable, normal, top-suspended pendulum consisting of a rod with a length of $2l$ having its center of mass at the midpoint is given by the expression

$$t = 2\pi \sqrt{\frac{I}{Mgl}} \rightarrow t = 2\pi \sqrt{\frac{4l}{3g}}. \qquad (T11.7)$$

When the moment of inertia for a uniform rod about its end point, $I = 4Ml^2/3$, is inserted, the second expression given in Eq. (T11.7) follows. If all of the body mass were concentrated at a single point a distance l from the pivot point, the expression for t becomes

$$t = 2\pi \sqrt{\frac{l}{g}}. \qquad (T11.8)$$

When the pendulum is inverted so that it stands vertically on its pivoted end, it inevitably starts to fall over. A measure of the time for this falling process is given by $t/4$ from Eq. (T11.8). Thus, for a pendulum of 1 m in total length, $l = 0.5$ m, and $t = 1.64$ s; if $l = 0.125$ m, then $t = 0.82$ s. The corresponding time for these rods to fall, $t/4$, would be 0.41 and 0.20 s, respectively.

Now, extend these results to consider a person's reaction time, which is less than 1 s but not much less than 0.5 s. Thus a person can balance a yardstick before it falls because he can respond to its unstable motion before the half a second or so that it takes to fall; a person cannot so easily balance a pencil, which falls in less than half that time. So, too, a woman standing on one ski can balance herself by attending to the biomechanical pressure feedback from her foot, which alerts her to her unbalanced condition and prompts her to bend her body to generate a reaction force at the support point in a manner similar to the way we move our palm to balance a yardstick. For example, the torque required to swing an arm generates a reaction torque that will rotate the rest of the body. Because the skier, as an inverted pendulum, is pinned at the foot by the ski, a horizontal force appears at that point that generates the external force necessary to move the center of mass and rotate the body as needed to regain equilibrium.

We can use Eq. (T11.7) to calculate the response time, $t/4$, needed for a person 6 ft tall to adjust to disequilibrium when skiing with the feet together in an attitude appropriately modeled by an inverted pendulum. That response time is about 0.78 s, well within the range of adequate response for most adults. For a child who is only 2 ft tall, however, the same response time, $t/4$, is much shorter: 0.45 s. Most children have not developed a reaction time fast enough to allow them to adjust their balance and maneuver successfully in such a short time. This is why small children, although they otherwise ski quite well, usually must ski with their feet somewhat apart; they simply cannot stay standing otherwise. Note that in these considerations the mass of the skier does not enter our discussion. The height of the skier and, to a much smaller extent, the skier's girth, affect the reaction time requirement.

Let us now set up a model of an operational system that simulates the body's response to the forces generated in a ski turn. In the steady-state turn, a bank angle is determined by the condition that the vector sum of all forces acting at the skier's center of mass, including the centrifugal inertial force and the gravitational force, must pass through the pivot point in the snow. That is, the torque acting on the skier produced by all forces is zero when the body is at the equilibrium tilt angle. Look again at the pendulum illustrated in Fig. T11.1. If the skier is on a horizontal plane, the vertical reaction force $R=Mg$ and the horizontal force F perpendicular to the motion must sum to pass through the center of mass so that no torques act on the skier. For the equilibrium bank angle γ_0 the force needed is $F=Mg \tan \gamma_0 \cong k\gamma_0$. The actual values of F and γ are then deviations from the equilibrium values.

An operational diagram of a structural control system that simulates what happens during a ski turn is shown in Fig. T11.2. Notice that $\Delta\gamma = \gamma - \gamma_0$ and $\Delta F = F - F_0$; the actual bank angle is γ and γ_0 is the desired equilibrium angle that the control system strives to attain. The block S, together with the comparison junction Σ, is the skier's neuromuscular system, which generates the body configuration needed to compensate for the out-of-balance

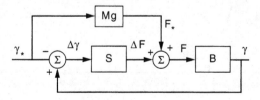

FIGURE T11.2. Structural control system diagram of a ski turn. [Reprinted with permission from J. M. Morawski, "Control Systems Approach to a Ski-Turn Analysis," J. Biomech. **6**, 267 (1973) (Elsevier Science Ltd., Oxford, England).]

body tilt, $\Delta\gamma$, by generating the force increment, ΔF, that will restore equilibrium to the system. The block B is the dynamic response of the body given by Eq. (T11.5). The actual trajectories of the skier's center of mass and feet must be simultaneously calculated as the turn progresses, so the overall block diagram is much more complicated. The independent variable for all outputs is the real time t.

Let us consider a case in which a skier follows a straight track but is off balance by some initial angle γ_0. The skier maneuvers to balance himself, but he overcorrects so that he tilts in the opposite direction by the angle γ—in effect, the skier oscillates back and forth without damping. The skier's velocity v is constant, $k>Mg$, and $k_1=0$. The distance moved along a straight reference line is designated by L; D and D_1 are displacements from the reference line of the skier's center of mass and feet, respectively.

As an initial condition, take $\gamma_0=30°$. The skier's velocity is 10 m/s; the moment of inertia, $I=13.7$ kg m^2; the height of the skier's center of mass from the pivot, l, equals 1 m; $k/Mg=2$; and the skier's mass M equals 70 kg. Figure T11.3 shows solutions for F and γ plotted as functions of time using these values.

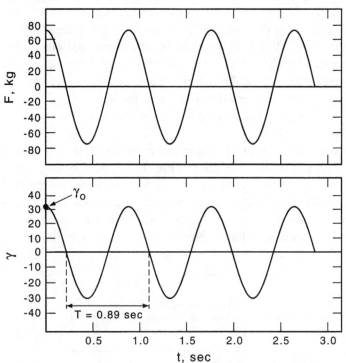

FIGURE T11.3. Time solution for F and γ for undamped oscillation. [Reprinted with permission from J. M. Morawski, "Control Systems Approach to a Ski-Turn Analysis," J. Biomech. **6**, 267 (1973) (Elsevier Science Ltd., Oxford, England).]

The tilt angle γ and the lateral force exerted on the ski by the snow vary over time in exactly the same manner because $F = k\gamma$. Because there is no damping and the motion starts with γ equal to 30°, the tilt angle oscillates back and forth by that amount. Other solutions for $D(L)$, the CM motion, using the same parameters but with different values of k/Mg as functions of L are shown in Fig. T11.4. The larger values of k/Mg represent stronger corrective responses by the skier. The period is given by expression (T11.4). Thus the frequency varies as $\sqrt{k/Mg - 1}$, so the wavelength changes by a factor of 2 going from $k/Mg = 1.5–3$. The amplitude of oscillation of the center of mass decreases as k increases; the amplitude of the ski oscillation is about 0.5 m. Figure T11.5 shows the displacement from the straight reference line $D(L)$ for the center of mass and for $D_1(L)$ for the skis. The amplitude $D(L)$ of the CM decreases markedly as the frequency of the oscillation increases. The curvature of the ski track generates the lateral force F needed to correct the large tilt angles of about 30° illustrated here.

Using this model to look at the mechanism and time history for initiating a turn is interesting. Actual skiers, of course, can perform an optimally damped maneuver; that is, they can correct their tilt angle to one side in such a way that they achieve their desired vertical stance rather than tilting over in the opposite direction. This implies that skiers use the rate-dependent term in F, that is, $k_1\gamma$, while skiing. Suppose the skier is initially at $\gamma = 0$ and decides to turn at $t = 0$ in a radius so that $\gamma_0 = 30°$ is required. Figure T11.6 shows the applied lateral force F and the decrement from the

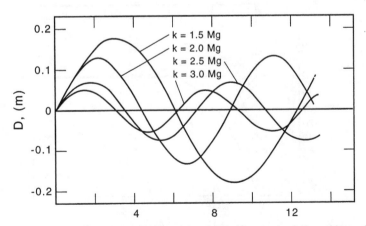

FIGURE T11.4. Space solutions for the lateral displacement of the center of mass trajectories, $D(L)$, as a function of distance down the fall line, L. [Reprinted with permission from J. M. Morawski, "Control Systems Approach to a Ski-Turn Analysis," J. Biomech. **6**, 267 (1973) (Elsevier Science Ltd., Oxford, England).]

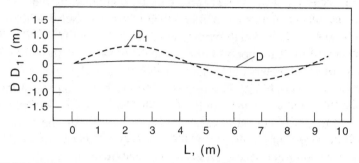

FIGURE T11.5. Space solutions for skier's center of mass, $D(L)$ (continuous line), and the skier's feet $D_1(L)$ (dashed line). [Reprinted with permission from J. M. Morawski, "Control Systems Approach to a Ski-Turn Analysis," J. Biomech. **6**, 267 (1973) (Elsevier Science Ltd., Oxford, England).]

equilibrium value F_0 for the case using the same values we used above in Fig. T11.3 with a value of k_1 for optimal damping.

No force toward the center of curvature may be applied until the skis are positioned outside the desired trajectory of the center of mass. Thus a

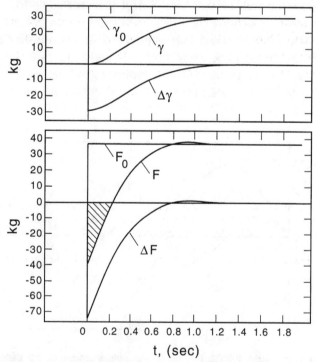

FIGURE T11.6. Time history of γ and F for a turn initiated at $t=0$ s. [Reprinted with permission from J. M. Morawski, "Control Systems Approach to a Ski-Turn Analysis," J. Biomech. **6**, 267 (1973) (Elsevier Science Ltd., Oxford, England).]

negative force to the outside of the curve must be applied to rotate the body to tilt to the inside of the curve and toward the desired tilt angle of γ_0. Recall Fig. 6.4, p. 136, which shows an example of the trajectories of the skier's center of mass and skis in this same maneuver. The skis curve to the outside for about 0.3 s, or 3 m of travel, before the skier's center of mass starts to curve toward the desired trajectory of the turn. Figure T11.6 above shows the tilt angle and the force needed for an optimally damped case to tilt a skier to $\gamma_0 = 30°$. In this case the value of k_1 is about optimum, since there is little over correction. The maneuver requires about 0.8 s or 8 m to progress from initiation into a steady-state condition.

Suppose that a turn of 60° is desired. The distance traveled over a turn where the force F balances the inertial centripetal force is 18.5 m and requires 1.8 s. At the completion of the turn, the skier must execute the reverse transition, righting his body to a vertical stance. The skis have to be pushed to the outside of the turn (or at least they must be set to skid) so that a force in excess of the centrifugal force is generated that tilts the skier's body upright as the transition is made to have the center of mass again on a straight path. About 0.8 s is needed for this maneuver, so about as much time is needed to get into and out of the turn as is needed during the course of the turn itself: 1.6 s is needed to get into and out of the turn; 1.8 s is needed to turn. Thus the skier must anticipate the maneuvers involved in this turn in order to perform them in this time frame. These results illustrate the advantage of using lateral projection, pushing off of the snow with the ski, to position the ski to turn. Using lateral projection, it takes almost no time to position the ski to turn, compared to the 0.8 s it takes to carve the skis into a turn.

REFERENCE

1. This discussion is indebted to J. M. Morawski's article, "Control Systems Approach to a Ski-Turn Analysis," J. Biomech. **6**, 267 (1973). Morawski uses γ to refer to the tilt angle in his "skier as pendulum" discussion, and we use Morawski's notation for the tilt angle here. Readers should note that in our preceding discussions of turning on skis in the main body of the book and in other Technotes, this tilt angle has been noted as ϕ.

TECHNOTE 12

SKI FLEXURE IN UNCOMPACTED SNOW

Let us consider the load distribution for a ski as it rides through uncompacted snow and compacts the snow bed [1]. The relation for the compaction pressure on a plunger pushing into the snow surface is $p = ky^n$, where p is the pressure needed to compact the snow to a depth y and k and n are functions of the initial density of the snow. The ski, however, does not act simply as a plunger; it slides forward at the same time. As a result, a plowing force acts upon the ski opposite to the ski's direction of motion. The snow-compaction loading over the base of the ski causes a reversed camber in the ski, and the ski assumes a configuration like the one shown in Fig. T12.1. The several forces and the geometry labeled in the figure will be referred to throughout the discussion that follows.

The reversed camber condition seen in the figure plays an important role when skiers carve turns in powder snow. The bowed shape of the ski lifts the tip of the ski up and out of the unpacked snow, and it determines the carving radius that the ski will make as it turns. Modeling how skis make parallel turns in soft snow depends on the radius of the ski's reversed camber and the lateral tilt of the ski as it tracks through the snow. The depth of the compaction track also determines the plowing force, which adds to the compaction force to retard the skier's forward motion. To carve turns effectively in unpacked snow, skiers must ride their skis in such a way that they generate the least compaction and plowing force while, at the same time, they generate an appropriate radius of reversed camber in their skis.

We can calculate the reversed camber, and hence the tip lift, for the ski shown in Fig. T12.1 above using the differential equation for determining the deflection of a beam that was given in Technote 2 as Eq. (T2.2), p. 200, and which we will rewrite here as

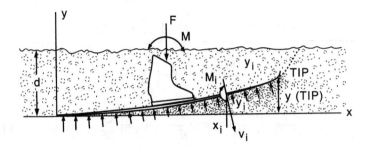

FIGURE T12.1. Configuration of a loaded ski compacting soft snow. The tail rides tangent to the track. The break in the ski shows the force and moment conventions. The desired parameters are the tip lift, y(tip), generated by the reversed camber of the ski, and the overall depth of the track. Note that the snow not compacted is plowed by the ski and the boot.

$$\frac{d^2 M}{dx^2} = w \rightarrow \frac{d^2}{dx^2}\left(B\,\frac{d^2 y}{dx^2}\right) = w. \qquad (\text{T12.1})$$

Recall that the bending stiffness B depends on x, the linear position on the beam, and w depends on both x and the beam deflection y. Recall also that $w = W(x)p(y)$, where $W(x)$ is the width and $p(y)$ is the pressure of the snow on the ski, which varies according to the deformation of the snow due to the loading applied. Readers who wish to calculate specific values of $p(y)$ should refer to relation (7.1) in Chap. 7, p. 160, and the discussion associated with it.

The problem before us is to solve the nonlinear differential equation (T12.1) with the compaction loading w. For convenience, y is measured upward from the compacted ski track, so $d - y$ is the actual compaction depth to be used in relation (T12.1) to compute a value for w. The horizontal distance x is measured along the track from the tail toward the tip in increments labeled x_1, x_2, and so on. The shear force V_i is the force applied to the forward portion of the ski at the position i. The same convention applies to the moment M_i. We assume an initial, constant compaction over the length of the ski that is set by the boot load F and use that to determine w_i in Eq. (T12.2). The forces applied at the ski bottom, w_i, are drawn to show the distribution of the bottom forces. We can solve this equation using stepwise integration by summing.

For the solution of the fourth-order differential equation before us, there must be four integrations, or stepwise summations, from one end to the other. These may be expressed as the following four relations:

$$w = -\frac{dV}{dx} \rightarrow V(x) = \left(\sum_{\text{tail}}^{x} w_1\right)\Delta x, \qquad (\text{T12.2})$$

$$V = -\frac{dM}{dx} \rightarrow M(x) = \left(\sum_{\text{tail}}^{x} V_i\right)\Delta x, \qquad (\text{T12.3})$$

$$\frac{M}{B} = \frac{d^2y}{dx^2} = \frac{dy'}{dx} \ \text{if} \ y' = \frac{dy}{dx}; \quad y' = \left[\sum_{\text{tail}}^{x}\left(\frac{M}{B}\right)_i\right]\Delta_x \qquad (\text{T12.4})$$

$$y' = \frac{dy}{dx} \rightarrow y(x) = \left(\sum_{\text{tail}}^{x} y_i'\right)\Delta x. \qquad (\text{T12.5})$$

To calculate the four sums from which the value of y at any point is generated, one must make the following initial assumptions. For the tail, $V=0$, $M=0$, $y'=0$, and $y=0$. For the tip, $V=0$, $M=0$, $y'=y'$(tip), and $y=y$(tip). The two boundary conditions, $V=0$ and $M=0$, that apply at the tip define the value of the boot force F and the position at which it must be applied. Remember the value of V_i at the position where F is applied changes appropriately. These boundary conditions are needed to generate a practical solution. The shear force V and the moment M must be zero at both the tip and the tail, because there are no forces exerted at the ends of the ski. The slope of the ski at a point on its tail, y_1', is zero because the tail slides directly in the track made by the forebody of the ski. The vertical displacement measured from the track, y, is zero because the origin is set at the track level. The applied load F must equal the sum of the reaction forces of the snow determined by the depth of compaction, $d-y(x)$. A moment M must also be applied, or the boot loading must be moved until the moment of all forces is zero. These values of F and M are just those needed to make V and M both zero at the tip.

Because ski and snow-compaction characteristics vary widely, we will work out only two representative cases in detail: one for a very soft, short ski, the Kneissl Ergo, and the other for a telemark ski, the Tua Excalibur. We will use compaction pressure functions derived from observations of uncompacted snow in which a mean compaction of about 30 cm was achieved for a 40 kg (88 lb) boot load.

The summation routine involves only simple sums and was programmed to be run by a simple spread sheet. A method of successive approximation was used as follows: The loading w_i was set constant at the value of the compaction depth d for the ski shown in Fig. T12.1 assuming the 40 kg load. Deflections were calculated from which the bottom reaction forces w_i were computed. The values for w_i were summed over the ski bottom to yield an opposite loading force F. The sum of all forces acting on the ski in the y direction is now zero, but the total moment about the origin is not zero. Move the location of F until the moment M at the tip also becomes zero. At this point the correct magnitude and location of the boot loading are determined because the boundary conditions are satisfied.

Next, the new set of y displacements is used to calculate another iteration of displacements. If this iteration differs only slightly from the previous set, the solution is satisfactory. For the cases tried, only one iteration was necessary. With each iteration, the value and location of the load force F may change, so a change in the track depth must be made to keep the total loading at the prescribed value. The desired parameter from this procedure is the deflection of the tip from the compacted ski track.

The curves of y_i are plotted in Fig. T12.2, which shows that some backward moment or displacement of the boot loading point toward the tail of the ski is needed to raise the tip of the ski. The three cases shown for the Tua ski (the full line) are (a) tail tangent to the track, (b) -70 cm, and (c) -30 cm tangent to the track with the remaining after body of the ski unloaded. For the Kneissl ski (illustrated as Δ points), only the cases for the tail and -30 cm tangent curves are shown. Each case represents a loading of 40 kg or 88 lb. These results agree with field experience: skiers must sit back on their skis to get their tips to ride up and generate minimum compaction.

For the cases in which the ski glided in the track or was unloaded for some distance forward from the tail, all of the boundary conditions shown above for the tail move up to the point of no loading or sliding without compaction. The points at -70 and -30 cm were used for these cases. The

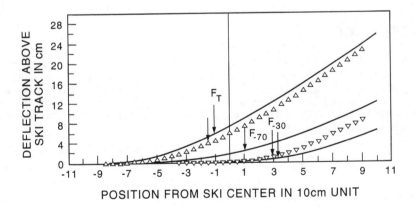

FIGURE T12.2. Flexure of a loaded ski on a deformable snow bed. The cases shown are for the Kneissl Ergo (triangular points) and the Tua Excalibur (full line) models. The displacement shown is measured from the compacted ski track.

sums were run just from that point forward. Loading at the tail of the ski generated a track depth of 30 cm; riding with the forward portion of the ski loaded produced a track depth of 35 cm for the same 40-kg load.

The important parameter for turning in unpacked snow is the tip lift of the ski, y(tip). When the load on the ski moves forward, the calculations show that tip lift decreases markedly, so the ski does not carve as easily. Furthermore, if the tip of the ski is not lifted to near the top of the snow, the ski will tend to dive because the snow does not compact enough to support the load of the skier. The skier must compensate for this tendency by moving the load on the ski back and tilting the tips of the skis up so that they ride out of the snow and generate the required compaction.

Snow conditions and ski properties vary widely, so only qualitative results serve to illustrate the points that may be drawn from these observations. In soft snow, skis with a large effective compaction or flotation area and a large tip lift are more desirable. A ski with a flexurally soft center and large shovel and tail widths will produce the needed flotation area and have enhanced tip lift. Short skis maneuver easily because their short length coupled with a large tip lift yields a much shorter carving radius in soft snow. We can roughly gauge the tip-lift value of a ski by supporting the ski at its tip and tail contact points, placing a load at about the center of the boot position, and then measuring the center deflection. Use a small weight for the load and scale the value to be roughly equal to the load of a skier's single leg load. The tip lift for a ski when its tail slides directly in the track made by the forebody of the ski will be about four times the corresponding bench measurement of the center deflection.

REFERENCE

1. For more information on how skis compact unpacked snow, see M. Mellor, "Properties of Snow," Monograph III-A1 (U.S. Army Corps of Engineers, Cold Regions Research and Engineering Laboratory, Hanover, NH, 1964), pp. 37–40. Where specific observations are noted in the discussion that follows, in most cases the data presented come from Mellor's work.

MELTWATER LUBRICATION

The meltwater film that lubricates the running surface of a ski once it begins to move with appreciable speed must arise from the frictional heat generated at the interface of the running surface and the snowpack [1]. The work done against the friction forces, whatever the nature of the forces themselves, represents an energy dissipation that is expressed in the form of heating. In addition to the heat produced by friction, some solar radiation that scatters within the snowpack and is absorbed by the ski base also contributes to the heating. The cumulative heat energy input heats the snow in the track and the ski itself until the interface of the contact surface of the ski and the snow comes to an effective melting temperature for the snow. The heat of fusion required to generate the meltwater holds the temperature rise to the melting point, 0 °C. Any additional available energy goes to increasing the generation of more liquid, not to raising the temperature above the melting point.

Because the ski passes continuously over a new snow surface, the heat that penetrates the snow is that which diffuses from the running surface of the ski at 0 °C into a snowpack having a uniform temperature of T_{sn}. The amount of heat energy depends on the total area of snow grains that make contact with the ski's running surface and on the duration for which any element of the snow's surface makes contact with the ski's running surface, or l/v, where l is the ski's length and v is the ski's velocity. We can write a heat balance equation that equates the input of heat energy from frictional work and solar absorption to the heat diffusing into the surface of the snow and into the ski that creates the lubricating film of meltwater:

$$Wv\mu + wl(1-\alpha)R = n\pi r^2 \rho_i L\dot{h} + q_i n\pi r^2 + wlq_{sl}. \qquad (T13.1)$$

Readers may recognize this equation as the mathematical expression that the cartoon skier drawn in Fig. 2.1, p. 12, holds in his head as he slides down the hill on a lubricating film of meltwater.

The first term on the left-hand side of the equation $(Wv\mu)$ represents the mechanical work done per unit of time or the power required to move the skier. This power is equivalent to the normal force F_N, which we assume to be equal to half of the skier's weight, times the coefficient of friction μ times the velocity v $(F_N\mu v)$. The second term $[wl(1-\alpha)R]$ is the expression for the rate of heating from solar radiation scattering through the snow and absorbed by the running surface of the ski. This heat value depends on the rectangular area of the total running surface of the ski, its width times its length, wl, and its absorptivity, $1-\alpha$, where α is the albedo of the running surface, times R, the solar radiation constant.

The first term on the right-hand side of the equation $(n\pi r^2 \rho_i L \dot{h})$ is the rate at which the cumulative heat from the left-hand side of the equation goes to create meltwater. This rate is given by the increase in the volume of the meltwater and the heat per unit volume required to melt that water. The number of ice grains contacting the ski base is given by n, so the effective area of the running surface of the ski that makes contact with the snow grains is modeled as a cylindrical space, $n\pi r^2$. The density of the ice grains is given by ρ_i; L is the heat of fusion, and, finally, the rate of growth of the meltwater thickness over the length of the ski base is $\dot{h}$. Multiplying these terms gives the rate of heating required to create the meltwater. The next term to the right $(q_i n\pi r^2)$ gives the heat conducted into the ice grains when an expression for the heat diffusion into the grains, q_i, is known from the thermal properties of ice and the initial temperature. The last term $(wl q_{sl})$ is the rate of heat conduction, q_{sl}, into the total running surface of the ski. In a steady state, q_{sl} is constant and depends on the thermal properties of the ski and its effective temperature, which is determined by the temperature of the snow and the ambient air.

Let us consider some hypothetical but appropriate values for the terms on the left-hand side of this equation to get some idea of the amounts of heat that might be generated by a ski as it goes down the slope. Using for $F_N=40$ kg, $v=10$ m/s, and $\mu=0.05$, the term $F_N\mu v$ yields 196 W for the rate of frictional heating. The solar constant R at normal incidence is 1340 W/m^2. Assume the absorptivity factor $(1-\alpha)$ is 0.9 and that the area of the single ski is 0.14 m^2. All of the solar radiation that penetrates the snow surface must on average scatter back out in all directions; if we assume that the available solar radiation is absorbed by the ski, then the heat generated by absorption of solar radiation becomes 1340×0.9×0.14, which equals 169 W. This represents an idealized maximum value; an actual value could easily be in the range of 100–120 W. The total heat generated is 196+169 W, or 365 W that would be spread over the running surface of the ski. For comparison, consider the heat given off by a 60-W light bulb or the heat generated by a small, 1200-W space heater when it is on a low setting. Clearly 365 W is ample heat to create a film of meltwater over the running surface of the ski.

This analysis shows us that the amount of heat that goes into the snow is large at the tip of the ski and decreases as the snow surface is heated. The heat going into the ski is reasonably constant, so the rate of production of the meltwater increases along the length of the running surface of the ski from the tip to the tail. At the tip of the ski, almost no meltwater is produced as the ski moves over new snow. The heat generated by the passing of the ski diffuses into the snow until the surface of the snow is heated to the melting point at some point back of the tip of the ski. As the snow under the ski is heated to a greater depth by heat diffusion, less heat is transmitted into the snow, and more liquid is produced. Thus this progressive production of liquid from the melting snow creates a thin, lubricating film of meltwater that extends from its minimum thickness at its origin some point back of the tip of the ski, to the ski's tail, where it rises to its maximum thickness and then leaves its mark in the glistening track the ski makes as it slides over the snow.

The analysis offered by this equational relation is idealized because it assumes that the loading on the ski, F_N, is uniform over the entire length of the ski, which, as we have seen, is not true in actual skiing. Nevertheless, the equation demonstrates that the tip, or shovel region, of the ski initiates the production of the lubricating film of meltwater, and this fact has had some effect on the manufacture of skis. Some skis claim to use base materials that increase their absorption of solar radiation, which presumably enhances the skis' ability to create a lubricating film of meltwater. Other skis claim to be fitted with base materials that offer a higher than usual coefficient of friction over their shovel regions. In this case, sacrificing some glide by allowing increased drag near the tip of the ski reaps an overall enhancement of glide from the more rapid heating of the snow that would occur over the length of the ski, bringing with that more rapid heating a more rapid growth of a lubricating film of meltwater. While there is some basis in the physics of skiing for the truth of these claims, wary skiers should always see for themselves how well the science that supports a ski's claim of improved performance translates into actual improved performance on the slope.

REFERENCE

1. This discussion is indebted to the work of S. C. Colbeck, "A Review of the Processes that Control Snow Friction," CRREL Monograph 92-2 (U.S. Army Corps of Engineers, Cold Regions Research and Engineering Laboratory, Hanover, NH, 1992).

GLOSSARY OF COMMONLY USED NOTATIONS AND SYMBOLS

Conventional usage dictates that some of the symbols listed below (for example, τ) may have different meanings in the context of different discussions. Wherever this occurs in the text, the symbol is defined in the context of the discussion. In this glossary, we cite our most common usage for notations and symbols.

W Gravitational weight of an object.

F_s Component of W in the plane of slope.

F_N Component of W normal to the plane of the slope.

F_p Component of gravitational force parallel to the ski.

F_{lat} Component of gravitational force in the plane of the slope and perpendicular to the ski.

F_{load} Total load imposed by the skier normal to the ski.

F_{tl} When making a turn, the total radial force parallel to the slope and normal to the path of motion, a component of F_{load}.

F_{reac} Total snow reaction force on the ski.

F_T Total snow reaction force on the ski in the plane of the slope, a component of F_{load}.

F_{treac} Total snow reaction force in the plane of the slope tangential to the path of motion of the skier's center of mass.

F_{lreac} Total snow reaction force in the plane of the slope lateral to the path of motion of the skier's center of mass.

F_f Slider friction force.

F_{plow} Slider drag force resulting from plowing forces.

F_{comp}	Slider drag force resulting from snow compaction forces.
F_D	Aerodynamic drag force.
F_L	Aerodynamic lift force.
F_I	Inertial force opposite in direction to acceleration.
F_C	Centrifugal inertial force.
F_{pl}	Force generated by poling.
X_{pl}	Chord length, or the projected length of the ski from tip to tail.
S	Shovel width.
W	Waist width.
T	Tail width.
C	Contact length: separation distance between the shovel width and the tail width.
SC	Sidecut, the separation between a line drawn from the shovel width and the tail width and the edge of the ski taken at the waist.
R_{SC}	Radius of the circular arc fitting points on the edge of the ski taken at the shovel width, waist, and the tail width.
α	Slope angle relative to a horizontal plane.
β	Ski traverse angle measured downslope from horizontal.
Δ	Tip lift in relation to a line drawn tangent to the tail of the ski.
ϕ	Tilt angle: angle of F_{load} relative to normal to the plane of the slope.
θ	Yaw angle between the skis when making a telemark turn; also the angle between the carving ski and the direction of motion.
Φ	Edge angle: the angle of the ski edge relative to the plane of the slope.
ψ	Effective tilt angle of both skis relative to the slope when the skis are tracking in soft snow.
μ	Coefficient of friction.
ρ	Density or mass per unit volume.
Y	Young's modulus: the tensile modulus of elasticity of a specific material.
M	Bending moment.
B	Bending stiffness coefficient for a section of a ski.
G	Torsional stiffness coefficient for a section of a ski.
V	Shear force for a section of a ski.
w	Loading per unit length on the ski bottom or edge.
p	Force per unit area or pressure.
d	Displacement.
v	Linear velocity.
a	Linear acceleration.
P	Linear momentum.
ω	Angular velocity.
Ω	Angular acceleration.

L	Angular momentum.
τ	Torque.
I	Moment of inertia for rotational motion.
T	Kinetic energy.
PE	Potential energy.

UNITS AND CONVERSIONS

Both English and metric or SI (standard international) units of measurement are used; data are usually presented as published without conversion.

Length	1 in.=2.54 cm, 1 m=39.37 in.=3.28 ft, 1 mile=1.609 km, 1 micron=1 μm=10^{-6}m
Force	This book follows a common practice of measuring force in units of weight, the force of gravity, in pounds or kilograms. The international standard unit is the Newton. Take acceleration of gravity, g=9.8 m/sec^2 =32 ft/sec^2. 1 kg=9.8 Newton, 1 N=0.225 lb.
Mass	1 kg=2.20 lb
Pressure	1 Pascal=1 N/m^2 1 kPa=10.2 grm/cm^2=0.145 lb/in.^{2} 1 lb/in.^{2}=70.5 grm/cm^2 1 Bar=10^6 dynes/cm^2=10^5 N/m^2=10^5 Pa 1 Atmosphere=1.0132 Bar=14.7 lb/in.^{2}=760 mm of Hg
Energy	1 Joule=1 Nm=0.738 ftlb 1 Cal=4.186 Joules=3.087 ftlbs Human biomechanical oxygen consumption for energy yields 1 liter O_2=5.05 Kcal=21.2 KJ=15.58 ftlbs
Temperature Kelvin units	T(K)=T(C)+273.15
Molar gas constant	R=8.314 J/Mol °K
Boltzmann constant	k=1.381×10^{-23} J/°K
Latent heat of melting ice	L_{sl}=79.6 cal/grm at 0° C
Latent heat of vaporization	L_{sg}=676.6 cal/grm at 0 °C for ice
Latent heat of vaporization	L_{lg}=595.9 cal/grm at 0 °C, or 538.7 cal/grm at 100 °C for water
Specific heat of ice	C=0.504+0.00199T cal/grm °C T temperature in °C
Specific heat of water	1.0 cal/grm °C at 15 °C

Bibliography

The literature that treats the science of skiing and snow is both numerous and widely scattered through a variety of publications, not all of them readily available to an inquiring reader. The most accessible literature is published in the scholarly books and journals on these subjects that may be found in the libraries of major universities. Popular ski magazines also, from time to time, will augment their generally qualitative discussions of skiing and snow with references to technical sources. Less readily accessible are the many handbooks, pamphlets, and brochures published by ski equipment manufacturers and the several professional groups associated with all aspects of the sport of skiing and the businesses associated with operating ski facilities.

The short discussions that follow note some of the representative works from this large body of literature, choosing, for the most part, books and articles that have fairly extensive references themselves, so that readers who go to these sources will find a ready path to still more sources. Finally, the bibliography of authored sources offers citations for all of the works referred to in this book as well as a selected range of other works whose extensive references or bibliographical citations recommend them.

PHYSICS

The nonscientist reader who wishes to inquire more deeply into the technical content of this book should read more about basic physics. Almost any introductory college textbook on physics is a good place to start. The three textbooks by Hewitt, by Giancoli, and by Halliday, Resnick, and Walker listed in the bibliography of authored sources are used fairly commonly and are likely to be available at a college or university bookstore. For the reader who wants a substantially more sophisticated presentation of the foundational principles of physics, see Richard P. Feynman's discussion that appears in Vol. I of *The Feynman Lectures on Physics*. Finally, *Handbook of Physics*, edited by E. U. Condon and H. Odishaw, is a ready and complete source of information about the foundations of physics, but it is probably more suitable for readers who already have some training in a science or engineering discipline.

SNOW

The best one-volume source of information on snowfall formation and snow cover on the ground is probably *The Handbook of Snow*, edited by D. M. Gray and D. H. Male. In that volume, see especially "Snowfall Formation" by R. S. Schemenauer, M. D. Berry, and J. B. Maxwell, and "Physics and Properties of Snowcover," by E. J. Langham. The best overall source for avalanche study is *The Avalanche Handbook*, by D. McClung and P. Schaerer, which offers specific sections on the formation, deposition, and metamorphism of the ground-cover snow.

SKI MECHANICS

A good starting point for understanding the mechanics of ski equipment is B. Glenne's "Mechanics of Skis," in the *Handbook of Snow*, cited earlier. There is some discussion of the mechanics of ski technique in the article by R. Perla and B. Glenne, "Skiing," also in the *Handbook of Snow*; their main emphasis, however, is on friction and wax performance. An extensive treatment of both the design of ski equipment and the skiing techniques that take advantage of that equipment may be found in J. Howe's book, *Skiing Mechanics*. Another book, *Universal Ski Techniques*, by G. Twardokens, also relates equipment to technique and cites many sources, in particular several from the European literature. For a more technical appreciation of the elastic properties of skis and snowboards, the reader may refer to either S. Timoshenko and J. N. Goodier, *Theory of Elasticity* or to Condon and Odishaw, *Handbook of Physics*. Two recent discussions of this topic appear in pamphlets published by PSIA, the Professional Ski Instructors of America: J. Vagners, "A Ski Instructor's Guide to the Physics and Biomechanics of Skiing," and R. LeMaster, "Skiing: The Nuts and Bolts." Finally, the reader will find a section entitled "Technical Analysis" in G. Joubert's book, *Skiing An Art... A Technique*, in which he describes much of the technical basis for the art of skiing.

FRICTION AND WAXING

For a narrative discussion of friction on skis and wax performance, see the article by R. Perla and B. Glenne, "Skiing," in the *Handbook of Snow*. The most comprehensive collection of the technical sources available on this topic is offered by S. C. Colbeck in his "Bibliography on Snow and Ice Friction," published in 1993 by the U.S. Army Corps of Engineers, Cold Regions Research and Engineering Laboratory (CRREL). A year earlier Colbeck published a CRREL monograph on ice and snow friction, "A Review of the Processes that Control Snow Friction," in which he cites in the context of his discussion of the topic much of the work later collected in the bibliography. These reports are infrequently collected in libraries. Readers seeking copies should contact the CRREL directly.

BACKCOUNTRY SKIING

For a good, general discussion of backcountry skiing, see V. Bein's *Mountain Skiing*. Readers will find more information on special applications for backcountry skiing in P. Shelton's book *The Snow Skier's Bible*. Readers interested in avalanche safety, which is a very important part of skiing the backcountry, should refer to books by D. McClung and P. Schaerer, *The Avalanche Handbook* and by B. Armstrong and K. Williams, *The Avalanche Book*. Both books cite many sources from the extensive literature on this topic. Readers interested in ski mountaineering or adventure skiing should seek out background information about mountaineering in general. "Mountaineering: The Freedom of the Hills," published by the Mountaineers of Seattle, Washington is a good general guide.

BIOMECHANICS AND MOUNTAIN MEDICINE

For background information about exercise, health, and sports, readers should see two books, by G. A. Brooks and T. D. Fahey, *Exercise Physiology*, and by K. F. Wells and K. Luttgens, *Kinesiology*. For information regarding the coordination of the human body with ski equipment, see the Proceedings of the Ski Trauma and Safety Symposia that are published by the American Society for Testing and Materials (ASTM) of Philadelphia, PA. The papers presented at these symposia range over such topics as orthopedic medicine, biomechanics, and statistical studies of ski injury and safety, as well as technical aspects of ski, binding, and boot designs.

Most skiing is done in the mountains at relatively high altitudes, which makes *Medicine of Mountaineering*, edited by J. A. Wilkerson, and the book by C. S. Houston, *Going Higher*, good sources for discussions of the several varieties of altitude sickness and guidelines for maintaining good health at altitudes of 5000 ft and higher. Houston's book features an extensive bibliography on this subject.

HANDBOOKS, GUIDES, AND SYMPOSIA

The Professional Ski Instructors of America (PSIA), 103 Van Gordon Street, Lakewood, Colorado, publishes a variety of manuals, monographs, and video instruction kits on all aspects of skiing. The most comprehensive PSIA publication is *The Official American Ski Technique* revised ed. (Cowles, New York, 1970). Some of the materials available on more specific topics are listed below:

"Teaching Concepts ATM (American Teaching Method)," H. Abraham, 1980.

"Strategies for Teaching ATS (American Teaching System)," 1987.

"The American Teaching System: Alpine Skiing," 2nd ed., 1993.

"The American Teaching System: Nordic Skiing," 1995.

"The American Teaching System: Snowboard Skiing," 1993.

"A Ski Instructor's Guide to the Physics and Biomechanics of Skiing," J. Vagners, 1995.

"Skiing: The Nuts and Bolts," R. LeMaster, 1995.

PSIA also publishes a journal, *The Professional Skier*, which appears quarterly and offers articles on instruction techniques and technical aspects of skiing.

The International Ski Association (ISA), P.O. Box 5070, 1380 GB Weest, The Netherlands, sponsors the INTERSKI Congresses from which printed contributions occasionally appear in proceedings or in other documents.

The K2 Corporation, a division of Anthony Industries, Vashon, WA, publishes a variety of materials. "The Ski Handbook," available from K2, discusses ski design, the engineering materials used in the construction of skis, and the dynamic properties of skis. Several other papers that treat ski design are also available from the K2 Corporation. Two of their publications, "Load Distribution of Nonprismatic, Prestressed Skis," by B. Glenne, J. Vandergrift, and A. DeRocco (no date), and "Basic Stuff for Ski Technicians," by B. Glenne (1992), list a number of references to papers that discuss the construction of skis.

Finally, the technical catalogues provided by manufacturers that list ski and snowboard equipment available for sale usually contain a fair amount of technical information about these products.

BIBLIOGRAPHY OF AUTHORED AND EDITED SOURCES

Allen, E. J. B. (1993), *From Skisport to Skiing: One Hundred Years of an American Sport, 1840–1940* (University of Massachusetts Press, Amherst).

Ambach, W., and Mayr, B. (1981), "Ski Gliding and Water Film," Cold Regions Sci. Technol. **5**, 59–65.

Armstrong, B., and Williams, K. (1993), *The Avalanche Book* (Fulcrum, Golden, CO).

Bein, V. (1982), *Mountain Skiing* (The Mountaineers, Seattle, WA).

Bowden, F. P., and Tabor, D. (1964), *The Friction and Lubrication of Solids* (Clarendon, Oxford), Part II.

Bowen, E. (1963), *The Book of American Skiing* (Bonanza, New York).

Bowers, R. W., and Fox, E. L. (1992), *Sports Physiology* (Brown, Dubuque, IA).

Brooks, G. A., and Fahey, T. D. (1985), *Exercise Physiology* (Macmillan, New York).

Carbone, C. (1994), *Women Ski* (World Leisure, Boston, MA).

Chen, J., and Kevorkian, V. (1971), "Heat and Mass Transfer in Making Artificial Snow," Ind. Eng. Chem. Process Des. Develop. **10**, 75.

Colbeck, S. C. (1981), "Introduction to the Basic Thermodynamics of Cold Capillary Systems," CRREL Special Report No. 81-6 (U.S. Army Corps of Engineers, Cold Regions Research and Engineering Laboratory, Hanover, NH).

Colbeck, S. C. (1994), "An Error Analysis of the Techniques Used in the Measurement of High Speed Friction on Snow," Ann. Glaciol. **19**, 19.

Colbeck, S. C. (1992), "A Review of the Processes that Control Snow Friction," CRREL Monograph 92-2 (U.S. Army Corps of Engineers, Cold Regions Research and Engineering Laboratory, Hanover, NH).

Colbeck, S. C. (1993a), "Bibliography on Snow and Ice Friction," CRREL Special Report No. 93-6 (U.S. Army Corps of Engineers, Cold Regions Research and Engineering Laboratory, Hanover, NH).

Colbeck, S. C. (1993b), "Bottom Temperatures of Skating Skis on Snow," unpublished report (U.S. Army Corps of Engineers, Cold Regions Research and Engineering Laboratory, Hanover, NH).

Colbeck, S., Akitaya, E., Armstrong, R., Gubler, H., Lafeuille, J., Lied, K., McClung, D., and Morris, E. (1990), "International Classification for Seasonal Snow on the Ground" (International Commission for Snow and Ice, World Data Center for Glaciology, University of Colorado, Boulder, CO).

Condon, E. U., and Odishaw, H., editors (1967), *Handbook of Physics*, 2nd ed. (McGraw-Hill, New York).

Daffern, T. (1983), *Avalanche Safety for Skiers and Climbers* (Rocky Mountain Books, Calgary).

de Mestre, N. (1990), *The Mathematics of Projectiles in Sports* (Cambridge University Press, Cambridge, U.K.).

Dyer, J. L. (1890), *Snow-Shoe Itinerant.* (Cranston and Stowe, Cincinnati, OH), (reprinted 1975, Father Dyer United Methodist Church, Breckenridge, CO).

English, B. (1984), *Total Telemarking* (East River, Crested Butte, CO).

Ettlinger, C. F., Johnson, R. J., and Shealy, J. E. (1995), "A Method to Help Reduce the Risk of Serious Knee Sprains Incurred in Alpine Skiing," Am. J. of Sports Med. **23** (5), 531–537.

Fall, R., and Wolber, P. K. (1995), "Biochemistry of Bacterial Ice Nuclei," in *Biological Ice Nucleation and Its Applications*, edited by R. E. Lee, G. J. Warren, and L. V. Gusta (APS, St. Paul, MN), pp. 63–83.

Figueras, J. M., Escalas, F., Vidal, A., Morgenstern, R., Bulo, J. M., Merino, J. A., and Espadaler-Gamisans, J. M. (1987), "The Anterior Cruciate Ligament Injury in Skiers," in *Ski Trauma and Safety: Sixth International Symposium* ASTM STP 938 (ASTM, Philadelphia), pp. 55–60.

Feynman, R., Leighton, R., and Sands, M. (1963), *The Feynman Lectures on Physics* (Addison-Wesley, Cambridge, MA), Vol. I.

Frederick, E. C., and Street, G. M. (1988), "Nordic Ski Racing, Biomechanical and Technical Improvements in Cross-Country Skiing," Sci. Am. **258** (2), T20.

Giancoli, D. C. (1995), *Physics*, 4th ed. (Prentice-Hall, Englewood Cliffs, NJ).

Glenne, B. (1981), "Mechanics of Skis," in *The Handbook of Snow*, edited by D. M. Gray and D. H. Male (Pergamon, Toronto).

Glenne, B. (1987), "Sliding Friction and Boundary Lubrication of Snow," Trans. ASME J. Tribol. **109** (4), 616.

Glenne, B., and Larsson, O. (1987), "Mechanics of a Giant Slalom Turn," Professional Skier, Winter/3, 23.

Gray, D. M., and Male, D. H., editors (1981), *The Handbook of Snow* (Pergamon, Toronto).

Halliday, D., Resnick, R., and Walker, J. (1993), *Fundamentals of Physics*, 4th ed. (Wiley, New York).

Hewitt, P. (1992), *Conceptual Physics*, 2nd ed. (Addison Wesley, Cambridge, MA).

Hobbs, P. V. (1974), *Ice Physics* (Clarendon, Oxford).

Holden, M. S. (1988), "The Aerodynamics of Skiing; Technology of Winning." Sci. Am. **258** (2), T4.

Houston, C. S. (1987), *Going Higher* (Little Brown, Boston, MA).

Howe, J. (1983), *Skiing Mechanics* (Poudre, LaPorte, CO).

Johnson, A. T. (1991), *Biomechanics and Exercise Physiology* (Wiley, New York).

Joubert, G. (1980), *Skiing an Art... a Technique* (Poudre, LaPorte, CO).

Kinosita, K. (1971), *Scientific Study of Skiing in Japan* (papers in European languages) (Hitachi, Tokyo).

Kuo, C. Y., Louie, J. K., and Mote, C. D. (1985), "Control of Torsion and Bending of the Lower Extremity During Skiing," in *Ski Trauma and Safety: Fifth International Symposium*, ASTM STP 860 (ASTM, Philadelphia), pp. 91–109.

LaDuca, R. J., Rice, A. F., and Ward, P. J. (1995), "Applications of Biological Ice Nucleation in Spray-Ice Technology," in *Biological Ice Nucleation and Its Applications*, edited by R. E. Lee, G. J. Warren, and L. V. Gusta (APS, St. Paul, MN), pp. 337–350.

Langham, E. J. (1981), "Physics and Properties of Snowcover," in *The Handbook of Snow*, edited by D. M. Gray and D. H. Male (Pergamon, Toronto).

Lash, B. (1970), "The Story of Skiing," in *The Official American Ski Technique* (Cowles, New York), pp. 3–14.

Lee, R. E., Jr., Warren, G. J., and Gusta, L. V. (1995), *Biological Ice Nucleation and Its Applications* (APS, St. Paul, MN).

LeMaster, R. (1995), "Skiing: The Nuts and Bolts" (Professional Ski Instructors of America, Lakewood, CO).

Lieu, D. K., and Mote, C. D., Jr. (1985), "Mechanics of the Turning Snow Ski," in *Skiing Trauma and Safety: Fifth International Symposium*, ASTM STP 860 (ASTM, Philadelphia), pp. 117–140.

Lock, G. S. H. (1991), *The Growth and Decay of Ice; Studies in Polar Research* (Cambridge University Press, Cambridge, U.K.).

Lund, M., Gillen, R., and Bartlett, M., editors (1982), *The Ski Book* (Arbor House, New York).

Mason, B. J. (1971), *The Physics of Clouds*, 2nd ed. (Clarendon, Oxford).

Mathews, J., and Walker, R. L. (1964), *Mathematical Methods of Physics* (W. A. Benjamin, New York).

McClung, D., and Schaerer, P. A. (1993), *The Avalanche Handbook* (The Mountaineers, Seattle, WA).

Mellor, M. (1964), "Properties of Snow," Monograph III-A1 (U.S. Army Corps of Engineers, Cold Regions Research and Engineering Laboratory, Hanover, NH).

Mellor, M. (1977), "Engineering Properties of Snow," J. Glaciol. **19** (81), 15.

Millikan, C. B. (1941), *Aerodynamics of the Airplane* (Wiley, New York).

Morawski, J. M. (1973), "Control Systems Approach to a Ski-Turn Analysis," J. Biomech. **6**, 267.

Mote, Jr., C. D., and Louie, J. K. (1982), "Accelerations Induced by Body Motion in Snow Skiing," J. Sound Vibr. **88**, 107.

Nelson R. C., McNitt-Gray, J., and Smith, G. (1986), "Biomechanical Analysis of Skating Technique in Cross Country Skiing" (Final Report to the U.S. Olympic Committee, Colorado Springs, CO).

Perla, R., and Glenne, B. (1981), "Skiing," in *Handbook of Snow*, edited by D. M. Gray and D. H. Male (Pergamon, Toronto).

Perla, R., and Martinelli, M. (1979), *Avalanche Handbook*, Agricultural Handbook 489 (USDA Forest Service, Washington, D.C.).

Pizialli, R. L., and Mote, C. D., Jr. (1972), "The Snow Ski as a Dynamic System," J. Dyn. Syst. Meas. Control, Trans. ASME **94** (2), 133.

Quinn, J., and Mote, C. D., Jr. (1992), "A Model for the Turning Snow Ski," J. Biomech. **26** (6), 609.

Raine, A. E. (1970), "Aerodynamics of Skiing," Sci. J. **6** (3), 26.

Reinisch, G. (1991), "A Physical Theory of Alpine Ski Racing." Spektrum Sportwissenschaft **I**, 27.

Renshaw, A. A., and Mote, C. D., Jr. (1991), "A Model for the Turning Snow Ski," in *Skiing Trauma and Safety: Eighth International Symposium*, ASTM STP (ASTM, Philadelphia), p. 1104.

Roberts, Jr., C. C. (1987), "Numerical Modeling of the Transient Dynamics of a Skier While Gliding," in *Biomechanics of Sport: A 1987 Update*, edited by E. D. Rekow, J. G. Thacker, and A. G. Erdman

(American Society of Mechanical Engineers, New York).

Rogers, R. R. (1976), *A Short Course in Cloud Physics* (Pergamon, New York).

Rouse, H., and Howe, J. W. (1953), *Basic Mechanics of Fluids* (John Wiley, New York), p. 181.

Sanders, J. R. (1979), *The Anatomy of Skiing* (Random House Vintage Books, New York).

Schaerer, P. A. (1981), *Avalanches* (Pergamon, Toronto).

Schemenauer, R. S., Berry, M. O., and Maxwell, J. B. (1981), "Snowfall Formation," in *The Handbook of Snow*, edited by D. M. Gray and D. H. Male (Pergamon, Toronto).

Seligman, G. (1980), *Snow Structure and Ski Fields* (International Glaciological Society, Cambridge, U.K.).

Sharkey, B. J. (1984), *Training for Cross-Country Racing* (Human Kinetics, Champaign, IL).

Shealy, J. E., and Miller, D. A. (1991), "A Relative Analysis of Downhill and Cross-Country Ski Injuries," in *Ski Trauma and Safety: Eighth International Symposium*, ASTM STP 1104 (ASTM, Philadelphia), pp. 133–143.

Shelton, P. (1991), *The Snow Skier's Bible* (Doubleday, New York).

Shimbo, M. (1971), "Friction of Snow on Ski Soles, Unwaxed and Waxed," in *Scientific Study of Skiing in Japan*, edited by K. Kinosita (Hitachi, Tokyo).

Smith, G. A. (1990), "Biomechanics of Crosscountry Skiing," Sports Med. **9** (5), 273.

Stenmark, I. (1990), "Ski Technique in the 1990s," Snow Country, March, 18.

Street, G. M., and Tsui, P. (1987), "Composition of Glide Waxes Used in Cross Country Skiing" (Biomechanics Laboratory, Pennsylvania State University, College Station, PA).

Swinson, D. B. (1992), "Physics and Skiing," Phys. Teacher **30**, 458.

Swinson, D. B. (1994), "Physics and Snowboarding," Phys. Teacher **32**, 530.

Timoshenko, S., and Goodier, J. N. (1951), *Theory of Elasticity*, 2nd ed. (McGraw-Hill, New York).

Torgersen, L. (1983), *Good Glide* (Human Kinetics, Champaign, IL).

Twardokens, G. (1992), *Universal Ski Techniques* (Surprisingly Well, Reno, NV).

Vaage, J. (1982), "The Norse Started It All," in *The Ski Book*, edited by M. Lund, R. Gillen, and M. Bartlett (Arbor House, New York), pp. 194–198.

Vagners, J. A. (1995), "Ski Instructor's Guide to the Physics and Biomechanics of Skiing" (Professional Ski Instructors of America, Lakewood, CO).

Walker, C. L. (1982), "A Way of Life," in *The Ski Book*, edited by M. Lund, R. Gillen, and M. Bartlett (Arbor House, New York), pp. 199–205.

Warren, G., and Wolber, P. K. (1991), "Molecular Aspects of Microbial Ice Nucleation," Mol. Microbiol. **5** (2), 239.

Warren, G. C., Colbeck, S. C., and Kennedy, F. E. (1989), "Thermal Response of Downhill Skis," Report No. 89–23 (U.S. Army Corps of Engineers, Cold Regions Research and Engineering Laboratory, Hanover, NH).

Wells, K. F., and Luttgens, K. (1976), *Kinesiology* (W. B. Sanders, Philadelphia).

West, J. B. (1984), "Human Physiology at Extreme Altitudes on Mount Everest," Science **223**, 24 February, 784.

Wilkerson, J. A. (1992), *Medicine for Mountaineering* (The Mountaineers, Seattle, WA).

Witherell, W. (1972), *How the Racers Ski* (W. W. Norton, New York).

Witherell, W., and Evrard, D. (1993), *The Athletic Skier* (Athletic Skier, Salt Lake City, UT).

Wolber, P. K. (1993), "Bacterial Ice Nucleation," Adv. Microbial Physiol. **34**, 203.

Yacenda, J. (1987), *High Performance Skiing* (Leisure, Champaign, IL).

Young, L. R., and Lee, S. M. (1991), "Alpine Injury Pattern at Waterville Valley, 1989 Update," in *Ski Trauma and Safety: Eighth International Symposium*, ASTM STP 1104 (ASTM, Philadelphia), pp. 125–132.

SUBJECT INDEX